NATEF Standards Job Sheets

Heating and Air Conditioning

(Test A7)

Jack Erjavec

THOMSON

DELMAR LEARNING

Australia   Canada   Mexico   Singapore   Spain   United Kingdom   United States

THOMSON

DELMAR LEARNING

NATEF Standards Job Sheets

Automotive Heating and
Air Conditioning
(Test A7)

Jack Erjavec

Business Unit Director:
Alar Elken

Executive Editor:
Sandy Clark

Acquisitions Editor:
Sanjeev Rao

Team Assistant:
Matthew Seeley

Executive Marketing Manager:
Maura Theriault

Marketing Coordinator:
Brian McGrath

Channel Manager:
Fair Huntoon

Executive Production Manager:
Mary Ellen Black

Production Manager:
Larry Main

Production Editor:
Betsy Hough

Cover Design:
Michael Egan

For more information contact Delmar Learning, Executive Woods, Maxwell Drive, Clifton Park, NY 12065-2919. Or find us on the World Wide Web at http://www.delmarlearning.com.

ISBN: 0-7668-6373-5

NOTICE TO THE READER

CONTENTS

PREFACE

With every passing day it gets harder to learn all that it takes to become a competent automotive technician. Technological advancements have allowed automobile manufacturers to build safer, more reliable, and more efficient vehicles. This is great for consumers, but along with each advancement comes a need for more knowledge.

Fortunately, students don't need to know it all. In fact, no one person knows everything about everything in an automobile. Although they don't know everything, good technicians do have a solid base of knowledge and skills. The purpose of this book is giving students a chance to develop all the skills and gain the knowledge of a competent technician. It is also the purpose of the guidelines established by the National Automotive Technicians Education Foundation (NATEF).

At the expense of much time and the work of many minds, NATEF has assembled a list of basic tasks for each of its certification areas. These tasks identify the basic skills and knowledge levels of competent technicians. The tasks also identify what is required for a student to start a successful career as a technician.

Most of what this book contains is job sheets. These job sheets relate to the tasks specified by NATEF. The main considerations in the creation of these job sheets were student learning and program certification by NATEF. Students are guided through standard industry-accepted procedures. While they are progressing, students are asked to report their findings and offer their thoughts on the steps they have just completed. The questions asked of the students are thought provoking and require students to apply what they know to what they observe.

The job sheets were designed to be generic; that is, whenever possible, the tasks can be performed on any vehicle from any manufacturer. Completion of the sheets does not require the use of specific brands of tools and equipment; rather, students use what is available. In addition, the job sheets can be used as a supplement to any good textbook.

Words to the Instructor   I suggest you grade these job sheets on completion and reasoning. Make sure the students answer all questions, and then look at the reasons to see if the task was actually completed and to get a feel for their understanding of the topic. It'll be easy for students to copy others' measurements and findings, but each student should have his or her own base of understanding, and that will be reflected in the explanations given.

Words to the Student   While completing the job sheets, you have a chance to develop the skills you need to be successful. When asked for your thoughts or opinions, think about what you observed. Think about what could have caused those results or conditions. You are not being asked to give accurate explanations for everything you do or everything you observe; you are only asked to think. Thinking leads to understanding. Good technicians are good because they have a basic understanding of what they are doing and of what they are doing it to.

NOTICE: SOME PARTS OF THIS COPY ARE DIFFERENT THAN THE PREVIOUS PRINTING OF THIS BOOK. THE CONTENTS HAVE BEEN UPDATED IN RESPONSE TO THE RECENT CHANGES MADE BY NATEF. *SEE PAGE 103 FOR DETAILS.*

HEATING AND AIR CONDITIONING SYSTEMS

To prepare you to learn what you should learn from completing the job sheets, some basics must be covered. This discussion begins with an overview of heating and air conditioning systems. Emphasis is placed on what they do and how they work, including the major components of heating and air conditioning systems and their role in the efficient operation of heating and air conditioning systems of all designs.

Preparation for work on an automobile would not be complete without addressing certain safety issues. This discussion covers what you should and should not do while working on heating and air conditioning systems, including the proper ways to deal with hazardous and toxic materials.

NATEF's task list for Heating and Air Conditioning certification is given, with definitions of some terms used to describe the tasks. This list gives you a good look at what the experts say you need to know before you can be considered competent to work on heating and air conditioning systems.

Following the task list are descriptions of the various tools and types of equipment you must be familiar with. These are the tools you will use to complete the job sheets. They are also the tools NATEF has identified as necessary for servicing heating and air conditioning systems.

After the tool discussion is a cross-reference guide that shows which NATEF tasks are related to specific job sheets. In most cases, there are single job sheets for each task. Some tasks are part of a procedure, in which case one job sheet may cover two or more tasks. The remainder of the book contains the job sheets.

HEATING SYSTEMS

The main components of an automotive heating system are the heater core, the heater control valve, the blower motor and the fan, and the heater and defroster ducts. The heating system works with the engine's cooling system and converts the heat from the coolant circulating inside the engine to hot air, which is blown into the passenger compartment. A heater hose transfers hot coolant from the engine to the heater control valve and from there to the heater core inlet. As the coolant circulates through the core, heat is transferred from the coolant to the tubes and fins of the core. Air blown through the core by the blower motor and fan then picks up heat from the surfaces of the core and transfers it into the passenger compartment. After giving up its heat, the coolant is then pumped out through the heater core outlet, where it is returned to the engine's cooling system to be heated again.

The heater control valve (also called the water flow valve) controls the flow of coolant into the heater core from the engine. In a closed position, the valve allows no flow of hot coolant to the heater core, thereby keeping the heater core cool. In an open position, the valve allows heated coolant to circulate through the heater core, thereby maximizing heater efficiency.

Cable-operated heater control valves are controlled directly by the heater control lever on the dashboard. Thermostatically controlled valves have a liquid-filled capillary tube located in the discharge air stream off the heater core. This tube senses air temperature, and the valve modulates the flow of water to maintain a constant temperature, regardless of engine speed or temperature.

Most heater control valves are vacuum operated. When a vacuum signal reaches the valve, a diaphragm inside the valve is raised, either opening or closing the valve against an opposing spring. When the temperature selection on the heater control panel is changed, vacuum to the valve is vented and the valve returns to its original position. Some vehicles don't use a heater control valve; instead, a heater door controls how much heat is released into the passenger compartment from the heater core.

The blower motor is usually located in the heater housing assembly. A multiposition switch on the control panel controls its speed. The switch works in connection with a stepped resistor block normally located near the heater housing. The resistor block controls blower motor speed. The typical resistor block has three or four resistors in series with the blower motor. The multiposition switch directs voltage to different locations in this series circuit. When low fan speed is selected, voltage is present at the beginning of the series circuit. High resistance causes slower operation of the blower motor. When high speed is selected, voltage is present at the end, or near the end, of the resistor assembly. Low resistance allows for higher fan speeds. On some vehicles, when the engine is running, the blower motor runs constantly at low speed.

Transferring heated air from the heater core to the passenger compartment is the job of the heater and defroster ducts. The ducts are typically part of a large plastic shell that connects to the necessary inside and outside vents. This ductwork also has mounting points for the evaporator and heater core assemblies. Also contained inside the duct are the doors required to direct air to the floor, dash, and/or windshield. Sometimes the duct is connected directly to the vents; other times, hoses are used.

Ventilation System

The ventilation system is designed to supply outside air to the passenger compartment through upper or lower vents, or both. Several systems are used to vent air into the passenger compartment. The most common is the flow-through system. In this arrangement, a supply of outside air, which is called *ram air,* flows into the car when it is moving. When the car is not moving, the heater fan can produce a steady flow of outside air. In operation, air is circulated throughout the passenger compartment. From there the air is forced outside the vehicle through an exhaust area.

AIR CONDITIONING SYSTEMS

It is important to know the basic theory of automotive air conditioning systems. Understanding how an A/C system moves heat from the confined space of the passenger compartment and dissipates it into the atmosphere will help you diagnose A/C problems.

An air conditioning system (Figure 1) is designed to pump heat from one point to another. All materials or substances, even those as cold as –459° Fahrenheit, have heat in them. Also, heat always flows from a warmer object to a colder one. The greater the temperature difference between the objects, the greater the amount of heat flow.

Objects can be in one of three forms or states: solid, liquid, or gas. When objects change from one state to another, large amounts of heat can be transferred. For example, when water temperature goes below 32°F, water changes from a liquid to a solid (ice). If the temperature of water is raised to 212°F, the liquid turns into a gas (steam). An interesting thing occurs when water, or any matter, changes from a solid to a liquid and then from a liquid to a gas. Additional heat is necessary to change the state of the substance, even though this heat does not register on a thermometer. For example, ice at 32°F requires heat to change into water, which will also be at 32°F. Additional heat raises the temperature of the water until it reaches the boiling point of 212°F. More heat is required to change water into steam. If the temperature of the steam were measured, however, it would also be 212°F. The amount of heat necessary to change the state of a substance is called *latent heat*—or *hidden heat*—because it cannot be measured with a thermometer. This hidden heat is the basic principle behind all air conditioning systems.

Pressure on a substance, such as a liquid, changes its boiling point. The greater the pressure on a liquid, the higher the boiling point. If pressure is placed on a vapor, the vapor condenses at a higher-

Evaporator

Service Valve
(Low-Pressure Side)

Compressor

Service Valve
(High-Pressure Side)

Receiver/ Dryer

Condenser

Figure 1 A typical air conditioning system. *Courtesy of American Honda Company.*

than-normal temperature. In addition, as the pressure on a substance is reduced, the boiling point can also be reduced. For example, the boiling point of water is 212°F. Increasing the pressure on the fluid can increase the boiling point. Reducing the pressure or placing the fluid in a vacuum can decrease it.

Refrigerants

The substance used to remove heat from the inside of an air-conditioned vehicle is called the refrigerant. On older automotive air conditioning systems, the refrigerant was Refrigerant-CFC-12 (commonly referred to as R-12 and Freon). R-12 is dichlorodifluoromethane (CCl_2F_2). By law, R-12 is no longer used in A/C systems. It was found that the chemicals in R-12 were damaging the earth's ozone layer, which is the earth's outermost shield of protection against the harmful effects of the sun's ultraviolet rays. The ozone depletion is caused, in part, by release of chlorofluorocarbons (CFCs) into the atmosphere. R-12 is in the chemical family of CFCs. Since air conditioning systems with R-12 are susceptible to leaks, further damage to the ozone layer could be avoided by stopping the use of R-12 in air conditioning units.

To replace R-12 in A/C systems, the manufacturers have agreed to use R-134a. This refrigerant may also be referred to as SUVA. R-134a is a hydrofluorocarbon (HFC) that causes less damage to the ozone layer when released to the atmosphere.

Although R-134a air conditioners operate in the same way and with the same basic components as R-12 systems, the two refrigerants are not interchangeable. Because it is less efficient than R-12, R-134a must operate at higher pressures. Therefore, R-134a systems use components designed for those higher pressures. To keep systems operating that were originally equipped with R-12, there are retrofit kits available for a changeover to R-134a. Although R-134a is less likely to have an adverse effect on the ozone layer, it still has the capability of contributing to the greenhouse effect when released into the air. Therefore, the recovery and recycling of R-12 and R-134a is mandatory by law. Laws have also been passed that dictate the certification of equipment and technicians.

Basic Operation

Refrigerants are used to carry heat from the inside of the vehicle to outside the vehicle. Automotive refrigerants have a low boiling point (the point at which evaporation occurs). As a refrigerant changes state, it absorbs a large amount of heat. Since the heat that it absorbs is from the inside of the vehicle, passengers become cooler.

To understand how a refrigerant is used to cool the interior of a vehicle, the effects of pressure and temperature on it must be understood. If the pressure of the refrigerant is high, so is its temperature. If the pressure is low, so is its temperature. Therefore, changing its pressure can change the temperature of a refrigerant.

To absorb heat, the temperature and pressure of the refrigerant is kept low. To dissipate heat, the temperature and pressure is kept high. As the refrigerant absorbs heat, it changes from a liquid to a vapor. As it dissipates heat, it changes from a vapor to a liquid. The change from a vapor to a liquid is called *condensation*. These two changes of state—evaporation and condensation—occur continuously as the refrigerant circulates through the air conditioning system.

An A/C system is a closed, pressurized system. It consists of a compressor, condenser, receiver/drier or accumulator, expansion valve or orifice tube, and an evaporator. In a basic air conditioning system, the heat is absorbed and transferred in the following steps.

1. Refrigerant leaves the compressor as a high-pressure, high-temperature vapor.
2. By removing heat via the condenser, the vapor becomes a high-pressure, high-temperature liquid.
3. Moisture and contaminants are removed by the receiver/drier, where the cleaned refrigerant is stored until it is needed.
4. The expansion valve controls the flow of refrigerant into the evaporator.
5. Heat is absorbed from the air inside the passenger compartment by the low-pressure, warm refrigerant, causing the liquid to vaporize and greatly decrease its temperature.
6. The refrigerant returns to the compressor as a low-pressure, low-temperature vapor.

The best way to understand the purpose of an A/C system's major components is to divide the systems into two sides: the high side and the low side. The high side is under high pressure and high temperature. The low side is the low-pressure, low-temperature side of the system.

The Compressor

The compressor separates the high-pressure and low-pressure sides of the system. Its primary purpose is to draw the low-pressure and low-temperature vapor from the evaporator and compress this vapor into high-temperature, high-pressure vapor. This action results in the refrigerant having a higher temperature than surrounding air and enables the condenser to condense the vapor back to a liquid. The secondary purpose of the compressor is to circulate or pump the refrigerant through the A/C system under the different pressures required for proper operation. The compressor is located on the engine and is driven by the engine's crankshaft via a drive belt.

Although many types of compressors are in use today, they are usually one of three types. A *piston compressor* can have its pistons arranged in an in-line, axial, radial, or V design. It is designed to have an intake stroke and a compression stroke for each cylinder. On the intake stroke, the refrigerant from the low side of the system (evaporator) is drawn into the compressor. Refrigerant is taken in through intake reed valves, which are one-way valves that control the flow of refrigerant vapors into the cylinder. During the compression stroke, the vaporous refrigerant is compressed. This increases both the pressure and the temperature of the heat-carrying refrigerant. The outlet or discharge-side reed valves then open to allow the refrigerant to move to the condenser. The outlet reed valves are the beginning of the high side of the system.

A common variation of a piston-type compressor is the variable displacement compressor. These compressors not only act as a regular compressor, but they also control the amount of refrigerant that passes through the evaporator. The pistons are connected to a wobble-plate. The angle of the wobble-plate determines the stroke of the pistons and is controlled by the difference in pressure between the compressor's outlet and inlet. When the stroke of the pistons increases, more refrigerant is being pumped and there is increased cooling.

A *rotary vane compressor* doesn't have pistons. Instead, it consists of a rotor with several vanes and a carefully shaped housing. As the compressor shaft rotates, the vanes and housing form chambers. The refrigerant is drawn through the suction port into these chambers, which become smaller as the rotor turns. The discharge port is located at the point where the gas is completely compressed.

The *scroll-type compressor* has a movable scroll and a fixed scroll that provide an eccentric-like motion. As the compressor's crankshaft rotates, the movable scroll forces the refrigerant against the fixed scroll and toward the center of the compressor. This motion pressurizes the refrigerant. The pressure of air moving in a circular pattern increases as it moves toward the center of the circle. A delivery port is positioned at the center of the compressor and allows the high-pressure refrigerant to flow into the air conditioning system.

Refrigerant Oils

Normally the only source of lubrication for a compressor is the oil mixed with the refrigerant. Because of the loads and speeds at which the compressor operates, proper lubrication is a must for long compressor life. The refrigerant oil required by the system depends on a number of things, but it is primarily dictated by the refrigerant used in the system. R-12 systems use a mineral oil. Mineral oil mixes well with R-12 without breaking down. Mineral oil, however, cannot be used with R-134a. R-134a systems require a synthetic oil, poly alkaline glycol (PAG). There are a number of different blends of PAG oil; always use the one recommended by the vehicle manufacturer or compressor manufacturer.

Compressor Clutches

Compressors are equipped with an electromagnetic clutch as part of the compressor pulley assembly. The clutch is designed to engage the pulley to the compressor shaft when the clutch coil is energized. The clutch provides a way for turning the compressor on or off.

The clutch is driven by power from the engine's crankshaft. When the clutch is engaged, power is transmitted from the pulley to the compressor shaft by the clutch drive plate. When the clutch is not engaged, the compressor shaft does not rotate, and the pulley freewheels.

The clutch is engaged by a magnetic field and disengaged by springs when the magnetic field is broken. When the controls call for compressor operation, the electrical circuit to the clutch is completed, the magnetic clutch is energized, and the clutch engages the compressor. When the electrical circuit is opened, the clutch disengages the compressor.

Condenser

The condenser consists of coiled refrigerant tubing mounted in a series of thin cooling fins to provide maximum heat transfer in a minimum amount of space. The condenser is normally mounted just in front of the vehicle's radiator. It receives the full flow of ram air from the movement of the vehicle or airflow from the radiator fan when the vehicle is standing still.

The condenser condenses or liquefies the high-pressure, high-temperature vapor coming from the compressor. To do so, it must give up its heat. The condenser receives very hot (normally 200° to 400°F), high-pressure refrigerant vapor from the compressor through its discharge hose. The refrigerant vapor enters the inlet at the top of the condenser, and as the hot vapor passes down through the condenser coils, heat moves from the hot refrigerant into the cooler air as it flows across the condenser coils and fins. This process causes a large quantity of heat to be transferred to the outside air and the refrigerant to change from a high-pressure hot vapor to a high-pressure warm liquid. This high-pressure warm liquid flows from the outlet at the bottom of the condenser through a line to the receiver/drier, or to the refrigerant metering device if an accumulator instead of a drier is used.

Receiver/Drier

The receiver/drier is a storage tank for the liquid refrigerant from the condenser. The refrigerant flows into the upper portion of the receiver tank, which contains a bag of desiccant (moisture-absorbing material, such as silica alumina or silica gel). The desiccant absorbs unwanted water and moisture in the refrigerant. As the refrigerant flows through the outlet at the lower portion of the receiver, a mesh screen filters it.

Included in many receiver/driers are a high-pressure fitting, a pressure relief valve, and a sight glass for determining the state and condition of the refrigerant in the system.

Accumulator

Most late-model systems are equipped with an accumulator rather than a receiver/drier. The accumulator is connected into the low side, at the outlet of the evaporator. The accumulator also contains a desiccant and is designed to store excess refrigerant and to filter and dry the refrigerant. If liquid refrigerant flows out of the evaporator, it will be collected by and stored in the accumulator. The main purpose of an accumulator is to prevent liquid from entering the compressor.

Thermostatic Expansion Valve or Orifice Tube

The refrigerant flow to the evaporator must be controlled to obtain maximum cooling, while ensuring complete evaporation of the liquid refrigerant within the evaporator. This is the job of a thermostatic expansion valve (TEV or TXV) or a fixed orifice tube.

The TEV is mounted at the inlet to the evaporator and separates the high-pressure side of the system from the low-pressure side. The TEV regulates refrigerant flow to the evaporator to prevent evaporator flooding or starving. In operation, the TEV regulates the refrigerant flow to the evaporator by balancing the inlet flow to the outlet temperature.

Both externally and internally equalized TEVs are used in air conditioning systems. The only difference between the two valves is that the external TEV uses an equalizer line connected to the evaporator outlet line as a means of sensing evaporator outlet pressure. The internal TEV senses evaporator inlet pressure through an internal equalizer passage. Both valves have a capillary tube to sense evaporator outlet temperature.

Like the TEV, the orifice tube is the dividing point between the high- and low-pressure parts of the system. However, its metering or flow rate control does not depend on comparing evaporator pressure and temperature. It is a fixed orifice. The flow rate is determined by pressure difference across the orifice and by subcooling. *Subcooling* is additional cooling of the refrigerant in the bottom of the condenser after it has changed from vapor to liquid. The flow rate through the orifice is more sensitive to subcooling than to pressure difference.

Evaporator

The evaporator, like the condenser, consists of a refrigerant coil mounted in a series of thin cooling fins. It provides a maximum amount of heat transfer in a minimum amount of space. The evaporator is usually located beneath the dashboard or instrument panel.

On receiving the low-pressure, low-temperature liquid refrigerant from the thermostatic expansion valve or orifice tube in the form of an atomized (or droplet) spray, the evaporator serves as a boiler or vaporizer. The regulated flow of refrigerant boils immediately. Heat from the core surface is lost to the boiling and vaporizing refrigerant, which is cooler than the core, thereby cooling the core. The air passing over the evaporator loses its heat to the cooler surface of the core, thereby cooling the air inside the car. As the process of heat loss from air to the evaporator core surface is taking place, any moisture (humidity) in the air condenses on the outside of the evaporator core and is drained off as water. This dehumidification of air adds to passenger comfort. It can also be used to control the fogging of windows.

Through the metering, or controlling, action of the thermostatic expansion valve or orifice tube, greater or lesser amounts of refrigerant are provided in the evaporator to adequately cool the car under all heat load conditions. If too much refrigerant is allowed to enter, the evaporator floods. This results in poor cooling due to the higher pressure (and temperature) of the refrigerant. The refrigerant can neither boil away rapidly nor vaporize. On the other hand, if too little refrigerant is metered, the evaporator starves. Poor cooling again results because the refrigerant boils away or vaporizes too quickly before passing through the evaporator.

Refrigerant Lines

All the major components of the system have inlet and outlet connections that accommodate either flare or O-ring fittings. There are three major refrigerant lines. Suction lines are located between the outlet side of the evaporator and the inlet side or suction side of the compressor. They carry the low-pressure, low-temperature refrigerant vapor to the compressor, where it again is recycled through the system. Suction lines are always distinguished from the discharge lines by touch and size. They are cold to the touch. The suction line is larger in diameter than the liquid line because refrigerant in a vapor state takes up more room than refrigerant in a liquid state.

The discharge or high-pressure line connects the compressor to the condenser. The liquid lines connect the condenser to the receiver/drier and the receiver/drier to the inlet side of the expansion valve. Through these lines, the refrigerant travels in its path from a gas state (compressor outlet) to a liquid state (condenser outlet) and then to the inlet side of the expansion valve, where it vaporizes on entry to the evaporator. Discharge and liquid lines are always very warm to the touch and easily distinguishable from the suction lines.

Aluminum tubing is commonly used to connect air conditioning components where flexibility is not required. R-134a systems are required to be fitted with quick-disconnect fittings through the system. These also have hoses specially made for R-134a. They have an additional layer of rubber that serves as a barrier to prevent the refrigerant from escaping through the pores of the hose.

Sight Glass

The sight glass allows the flow of refrigerant in the lines to be observed. A sight glass is normally found on systems using R-12 and a thermal expansion valve. It is not commonly found on R-134a systems. It can be located on the receiver/drier or in-line between the receiver/drier and the expansion valve or tube.

Accumulator systems do not have a receiver/drier to separate the gas from the liquid as it flows from the condenser. The liquid line always contains a certain amount of bubbles, therefore, it would be useless to have a sight glass in these systems. Pressure and performance testing are the only ways to identify low refrigerant levels.

Blower Motor and Fan

The blower motor/fan assembly is located in the evaporator housing. The blower, which is basically the same type as those used in heater systems, draws warm air from the passenger compartment, forces it over the coils and fins of the evaporator, and blows the cooled, cleaned, and dehumidified air into the passenger compartment. A fan switch controls the blower motor. During cold weather, the blower motor/fan assembly provides the airflow, and the heater core provides the heat for the passenger compartment.

Evaporator Pressure Control Systems

Evaporator controls maintain back pressure in the evaporator. Because of the refrigerant temperature-pressure relationship, the effect is to regulate evaporator temperature. The temperature is controlled to a point that provides effective air cooling but prevents the freezing of moisture that condenses on the evaporator.

In this type of system, the compressor operates continually when dash controls are in the air conditioning position. An evaporator pressure control valve automatically controls evaporator outlet air temperature. These types of valves throttle the flow of refrigerant out of the evaporator as required to establish a minimum evaporator pressure and thereby prevent freezing of condensation on the evaporator core.

Cycling Clutch Systems

In cycling clutch systems, the compressor is run intermittently by controlling the application and release of its clutch by a thermostatic or pressure switch. The thermostatic switch senses the evaporator's outlet air temperature through a capillary tube that is part of the switch assembly. With a high sensing temperature, the thermostatic switch is closed and the compressor clutch is energized. As the evaporator outlet temperature drops to a preset level, the thermostatic switch opens the circuit to the compressor clutch. The compressor then ceases to operate until such time as the evaporator temperature rises above the switch setting. The term *cycling clutch* derives from this on-and-off operation. In effect, the thermostatic switch is calibrated to allow the lowest possible evaporator outlet temperature that prevents the freezing of condensation on the evaporator.

Variations of the cycling clutch system include a system with a TEV and a system with an orifice tube.

Compressor Controls

There are many controls used to monitor and trigger the compressor during its operational cycle. Each of these represents the most common protective control devices designed to ensure safe and reliable operation of the compressor.

An ambient temperature switch senses outside air temperature and is designed to prevent compressor clutch engagement when air conditioning is not required or when compressor operation might cause internal damage to seals and other parts.

On some vehicles, the ambient switch is located in the air inlet duct of the air conditioning system or near the radiator. It is not required on systems with a thermostatic or pressure switch.

In cycling clutch systems, a thermostatic switch is placed in series with the compressor clutch circuit so it can turn the clutch on or off. It deenergizes the clutch and stops the compressor if the evaporator is at the freezing point. It also may control the air temperature by turning the compressor on and off intermittently.

When the temperature of the evaporator approaches the freezing point, the thermostatic switch opens the circuit and disengages the compressor clutch. The compressor remains inoperative until the evaporator temperature rises to the preset temperature, at which time the switch closes and compressor operation resumes.

A pressure cycling switch is connected in series with the compressor clutch. Like the thermostatic switch, the turning on and off of the pressure cycling switch controls the operation of the compressor.

The low-pressure cutoff or discharge pressure switch is located on the high side of the system and senses any low-pressure conditions. This switch is tied into the compressor clutch circuit, allowing it to immediately disengage the clutch when the pressure falls too low.

The high-pressure cutoff switch is normally located near the compressor and is also wired in series with the compressor clutch. This switch is designed to open and disengage the clutch at a specified high pressure. When the pressure drops below a specified pressure, it again closes and reengages the clutch.

A high-pressure relief valve may be installed on the receiver/drier, compressor, or elsewhere in the high side of the system. It is a high-pressure protection device that opens to bleed off any excessive pressure in the system.

The compressor control valve regulates the crankcase pressure in some compressors. It has a pressure-sensitive bellows assembly exposed to the suction side. This bellows acts on a ball-and-pin valve, which is exposed to high-side pressure. The bellows also controls a bleed port that is exposed to the low side. The control valve is continuously modulating, changing the displacement of the compressor according to pressure or temperature.

An electronic cycling clutch switch prevents evaporator freeze-up by sending a signal to the Powertrain Control Module (PCM). The PCM, in turn, cycles the compressor on and off by monitoring suction line temperature. If the temperature gets too low, the switch opens the input circuit to the computer, which causes the A/C clutch relay to open, which disengages the compressor clutch. Often this switch is the thermostatic switch at the evaporator.

Case/Duct Systems

The purpose of a heater/air conditioner/case and duct system is twofold. It is used to house the heater core and the evaporator and to direct the selected supply air through these components into the passenger compartment. The selected supply air can be either fresh (outside) or recirculated air, depending on the system mode. After the air is heated or cooled, it is delivered to the floor outlet, dash panel outlets, or the defrost outlets.

Manufacturers use two basic types of duct systems. The stacked core reheat system uses the heater control valve as the main control. All air enters the passenger compartment through the heater core. An access door controls only fresh or recirculated air. The blend air reheat system uses a mixture of heated and cooled air to provide comfort. The duct system brings in outside air and/or recirculated air across the heater core and evaporator. The door controls the source of the air entering into the passenger compartment.

Temperature Control Systems

Temperature control systems for air conditioners usually are connected with heater controls. Most heater and air conditioning systems use the same plenum chamber for air distribution. Two types of air conditioning controls are used: manual/semiautomatic and automatic.

Air conditioner manual/semiautomatic temperature controls operate in a manner similar to heater controls. Depending on the control setting, doors are opened and closed to direct airflow. The amount of cooling is controlled manually through the use of control settings and blower speed.

Automatic or electronic temperature control systems maintain a specific temperature automatically. To maintain a selected temperature, heat sensors send signals to a computer unit that controls compressor, heater valve, blower, and plenum door operation. A typical electronic control system might contain a coolant temperature sensor, in-car temperature sensor, outside temperature sensor, high-side temperature switch, low-side temperature switch, low-pressure switch, vehicle speed sensor, throttle position sensor, sunload sensor, and power steering cutout switch.

Three types of control panels may be found: manual, push-button, or touch pad. All serve the same purpose—they provide operator input control for the air conditioning and heating system. Some control panels have features that other panels do not have, such as provisions to display in-car and outside air temperature in degrees. Provisions are made on the control panel for operator selection of an in-car temperature between 65° and 85°F in one-degree increments. Some have an override feature that provides for a setting of either 60° or 90°F. Either of these two settings overrides all in-car temperature control circuits to provide maximum cooling or heating conditions.

SAFETY

In an automotive repair shop, there is great potential for serious accidents simply because of the nature of the business and the equipment used. Through carelessness, the automotive repair industry can be one of the most dangerous occupations. However, the chances of being injured while working on a car are close to nil if you learn to work safely and use common sense. Safety is the responsibility of everyone in the shop.

Personal Protection

Some procedures, such as grinding, result in tiny particles of metal and dust that are thrown off at very high speeds. These metal and dirt particles can easily get into your eyes, causing scratches or cuts on your eyeball. Pressurized gases and liquids escaping a ruptured or damaged hose fitting can spray a great distance. If these chemicals get into your eyes, they can cause blindness. This is especially true of refrigerants. Refrigerants can quickly freeze an eye and cause serious damage.

Eye protection should be worn whenever you are exposed to these risks. To be safe, you should wear safety glasses whenever you are working in the shop. Some procedures may require that you wear other eye protection in addition to safety glasses; for example, when cleaning parts with a pressurized spray, you should wear a face shield. The face shield not only gives added protection to your eyes but it also protects the rest of your face.

If chemicals such as battery acid, fuel, or solvents get into your eyes, flush them continuously with clean water. Have someone call a doctor and get medical help immediately.

Your clothing should be well fitted and comfortable but made with strong material. Loose, baggy clothing can easily get caught in moving parts and machinery. Some technicians prefer to wear coveralls or shop coats to protect their personal clothing. Your work clothing should offer you some protection but should not restrict your movement.

Long hair and loose, hanging jewelry can create the same type of hazard as loose-fitting clothing. They can get caught in moving engine parts and machinery. If you have long hair, tie it back or tuck it under a cap.

Never wear rings, watches, bracelets, or neck chains. These can easily get caught in moving parts and cause serious injury.

Always wear leather or similar material shoes or boots with non-slip soles. Steel-tipped safety shoes can give added protection to your feet. Jogging or basketball shoes, street shoes, or sandals are inappropriate in the shop.

Good hand protection is often overlooked. A scrape, cut, or burn can limit your effectiveness at work for many days. A well-fitted pair of heavy work gloves should be worn during operations such as grinding and welding or when handling high-temperature components. Always wear approved rubber gloves when handling strong and dangerous caustic chemicals.

When working on A/C systems, protect your hands from exposure to the refrigerant. Escaping refrigerant can cause frostbite to the area of skin exposed to the leak. In addition to wearing gloves, it is wise to use a rag when connecting and disconnecting service gauges to an A/C system.

Many technicians wear thin, surgical-type latex gloves whenever they are working on vehicles. These offer little protection against cuts but do offer protection against disease and grease buildup under and around your fingernails. These gloves are comfortable and are quite inexpensive.

Accidents can be prevented simply by the way you act. Following are some guidelines to follow while working in a shop. The list does not include everything you should or shouldn't do; it merely gives some things to think about.

- Never smoke while working on a vehicle or while working with any machine in the shop.
- Playing around is not fun when it sends someone to the hospital.
- To prevent serious burns, keep your skin away from hot metal parts, such as the radiator, exhaust manifold, tailpipe, catalytic converter, and muffler.
- Always disconnect electric engine cooling fans when working around the radiator. Many of these will turn on without warning and can easily chop off a finger or hand. Make sure you reconnect the fan after you have completed your repairs.
- When working with a hydraulic press, make sure the pressure is applied in a safe manner. It is generally wise to stand to the side when operating the press.
- Properly store all parts and tools by putting them away in a place where people will not trip over them. This practice not only cuts down on injuries, it also reduces time wasted looking for a misplaced part or tool.

Work Area Safety

Your entire work area should be kept clean and safe. Any oil, coolant, or grease on the floor can make it slippery. To clean up oil, use commercial oil absorbent. Keep all water off the floor. Water is slippery on smooth floors, and electricity flows well through water. Aisles and walkways should be kept clean and wide enough to move through easily. Make sure the work areas around machines are large enough to safely operate the machine.

Gasoline is a highly flammable volatile liquid. Something that is *flammable* catches fire and burns easily. A *volatile* liquid is one that vaporizes very quickly. *Flammable volatile* liquids are potential firebombs. Always keep gasoline or diesel fuel in an approved safety can, and never use gasoline to clean your hands or tools.

Handle all solvents (or any liquids) with care to avoid spillage. Keep all solvent containers closed, except when pouring. Proper ventilation is very important in areas where volatile solvents and chemicals are used. Solvent and other combustible materials must be stored in approved and designated storage cabinets or rooms with adequate ventilation. Never light matches or smoke near flammable solvents and chemicals, including battery acids.

Oily rags should also be stored in an approved metal container. When oily, greasy, or paint-soaked rags are left lying about or are not stored properly, they can spontaneously combust. Spontaneous combustion refers to fire that starts by itself, without a match.

Disconnecting the vehicle's battery before working on the electrical system, or before welding, can prevent fires caused by a vehicle's electrical system. To disconnect the battery, remove the negative or ground cable from the battery and position it away from the battery.

Know where all the shop's fire extinguishers are located. Fire extinguishers are clearly labeled as to type and types of fire they should be used on. Make sure you use the correct type of extinguisher for the type of fire you are dealing with. A multipurpose dry chemical fire extinguisher will put out ordinary combustibles, flammable liquids, and electrical fires. Never put water on a gasoline fire—the water will just spread the fire. The proper fire extinguisher smothers the flames.

During a fire, never open doors or windows unless it is absolutely necessary; the extra draft will only make the fire worse. Make sure the fire department is contacted before or during your attempt to extinguish a fire.

Tool and Equipment Safety

Careless use of simple hand tools, such as wrenches, screwdrivers, and hammers, causes many shop accidents that could be prevented. Keep all hand tools free of grease and in good condition. Tools that slip can cause cuts and bruises. If a tool slips and falls into a moving part, it can fly out and cause serious injury.

Use the proper tool for the job, and make sure the tool is of professional quality. Using poorly made tools or the wrong tools can damage parts, the tool itself, or you. Never use broken or damaged tools.

Safety around power tools is very important. Serious injury can result from carelessness. Always wear safety glasses when using power tools. If the tool is electrically powered, make sure it is properly grounded. Before using it, check the wiring for bare wires and for cracks in the insulation. When using electrical power tools, never stand on a wet or damp floor. Never leave a running power tool unattended.

When using compressed air, wear safety glasses or a face shield, or both. Particles of dirt and pieces of metal, blown by the high-pressure air, can penetrate your skin or get into your eyes.

Before using a compressed air tool, check all hose connections. Always hold an air nozzle or air control device securely when starting or shutting off the compressed air. A loose nozzle can whip suddenly and cause serious injury. Never point an air nozzle at anyone. Never use compressed air to blow dirt from your clothes or hair. Never use compressed air to clean the floor or workbench.

Always be careful when raising a vehicle on a lift or a hoist. Adapters and hoist plates must be positioned correctly to prevent damage to the vehicle's underbody. There are specific lift points that allow the weight of the vehicle to be evenly supported by the adapters or hoist plates. The correct lift points can be found in the vehicle's service manual. Before operating any lift or hoist, carefully read the operating manual and follow the operating instructions.

Once you know the lift supports are properly positioned under the vehicle, raise the lift until the supports contact the vehicle. Then, check the supports to make sure they are in full contact with the vehicle. Shake the vehicle to make sure it is securely balanced on the lift, and then raise the lift to the desired working height. Before working under a car, make sure the lift's locking devices are engaged.

A vehicle can be raised off the ground by a hydraulic jack. The jack's lifting pad must be positioned under an area of the vehicle's frame or at one of the manufacturer's recommended lift points. Never place the pad under the floor pan or under steering and suspension components, which are easily damaged by the weight of the vehicle. Always position the jack so the wheels of the vehicle can roll as the vehicle is being raised.

Safety stands, also called jack stands, should be placed under a sturdy chassis member, such as the frame or axle housing, to support the vehicle after it has raised by a jack. Once the safety stands are in position, the hydraulic pressure in the jack should be slowly released until the weight of the vehicle is on the stands. Never move under a vehicle when it is supported only by a hydraulic jack. Rest the vehicle on the safety stands before moving under the vehicle.

Cleaning parts is a necessary step in most repair procedures. Always wear the appropriate protection when using chemical, abrasive, and thermal cleaners.

Vehicle Operation

When the customer brings a vehicle in for service, certain driving rules should be followed to ensure your safety and the safety of those working around you. For example, before moving a car into the shop, buckle your safety belt. Make sure no one is near, the way is clear, and there are no tools or parts under the car before you start the engine.

Check the brakes before putting the vehicle in gear. Then, drive slowly and carefully in and around the shop.

If the engine must be running while work is done on the car, block the wheels to prevent the car from moving. Place the transmission in park for automatic transmissions or in neutral for manual transmissions. Set the parking (emergency) brake. Never stand directly in front of or behind a running vehicle.

Run the engine only in a well-ventilated area to avoid the danger of poisonous carbon monoxide (CO) in the engine exhaust. CO is an odorless but deadly gas. Most shops have an exhaust ventilation system; always use it. Connect the hose from the vehicle's tailpipe to the intake for the vent system. Make sure the vent system is turned on before running the engine. If the work area does not have an exhaust venting system, use a hose to direct the exhaust out of the building.

HAZARDOUS MATERIALS AND WASTES

A typical shop contains many potential health hazards for those working in it. These hazards can cause injury, sickness, impairment, discomfort, and even death. Here is a short list of the different classes of hazards.

- Chemical hazards are caused by high concentrations of vapors, gases, or solids in the form of dust.
- Hazardous wastes are substances that are the result of a service.
- Physical hazards include excessive noise, vibration, pressures, and temperatures.
- Ergonomic hazards are conditions that impede normal or proper body position and motion.

There are many government agencies charged with ensuring safe work environments for all workers. These include the Occupational Safety and Health Administration (OSHA), Mine Safety and Health Administration (MSHA), and National Institute for Occupational Safety and Health (NIOSH). These agencies, in addition to state and local governments, have instituted regulations that must be understood and followed. Everyone in a shop is responsible for adhering to these regulations.

An important part of a safe work environment is the employees' knowledge of potential hazards. Right-to-know laws concerning all chemicals protect every employee in the shop. The general intent of right-to-know laws is for employers to provide their employees with a safe working place as it relates to hazardous materials.

All employees must be trained about their rights under the legislation, the nature of the hazardous chemicals in their workplace, and the contents of the labels on the chemicals. All information about each chemical must be posted on Material Safety Data Sheets (MSDS) and must be accessible. The manufacturer of the chemical must give these sheets to its customers if it is asked to do so. They detail the chemical composition and precautionary information for all products that can present a health or safety hazard.

Employees must become familiar with the general uses, protective equipment, accident or spill procedures, and any other information regarding the safe handling of the hazardous material. This training must be given to employees annually and provided to new employees as part of their job orientation.

All hazardous material must be properly labeled, indicating what health, fire, or reactivity hazard it poses and what protective equipment is necessary when handling each chemical. The manufacturer of the hazardous materials must provide all warnings and precautionary information, which must be read and understood by the user before use. A list of all hazardous materials used in the shop must be posted for the employees to see.

Shops must maintain documentation on the hazardous chemicals in the workplace, proof of training programs, records of accidents or spill incidents, and satisfaction of employee requests for specific chemical information via the MSDS. A general right-to-know compliance procedure manual must be used in the shop.

When handling any hazardous materials or hazardous waste, make sure you follow the required procedures for handling such material. Wear the proper safety equipment listed on the MSDS, which includes the use of approved respirator equipment.

Some of the common hazardous materials that automotive technicians use are cleaning chemicals, fuels (gasoline and diesel), paints and thinners, battery electrolyte (acid), used engine oil, refrigerants, and engine coolant (antifreeze).

Many repair and service procedures generate what are known as hazardous wastes. Dirty solvents and cleaners are good examples of hazardous wastes. Something is classified as a hazardous waste if it is on the Environmental Protection Agency (EPA) list of known harmful materials or has one or more of the following characteristics.

- *Ignitability.* A liquid with a flash point below 140°F or a solid that can spontaneously ignite.
- *Corrosivity.* A substance dissolves metals and other materials or burns the skin.
- *Reactivity.* Any material that reacts violently with water or other materials or releases cyanide gas, hydrogen sulfide gas, or similar gases when exposed to low-pH acid solutions. This includes materials that generate toxic mists, fumes, vapors, or flammable gases.
- *Toxicity.* Materials that leach one or more of eight heavy metals in concentrations greater than 100 times primary drinking water standard concentrations.

Complete EPA lists of hazardous wastes can be found in the Code of Federal Regulations. It should be noted that no material is considered hazardous waste until the shop is finished using it and ready to dispose of it.

The following list describes the recommended procedure for dealing with some of the common hazardous wastes. Always follow these and any other mandated procedures.

Refrigerants   Recover or recycle refrigerants (or do both) during the service and disposal of motor vehicle air conditioner and refrigeration equipment. It is not allowable to knowingly vent refrigerants to the atmosphere. Recovering or recycling during servicing must be performed by an EPA-certified technician using certified equipment and following specified procedures.

Containers   Cap, label, cover, and properly store above-ground and outdoors any liquid containers and small tanks within a diked area and on a paved impermeable surface to prevent spills from running into surface or ground water.

Oil   Recycle oil. Set up equipment, such as a drip table or screen table with a used oil collection bucket, to collect oils dripping off parts. Place drip pans underneath vehicles that are leaking fluids onto the storage area. Do not mix other wastes with used oil, except as allowed by your recycler. Used oil generated by a shop (or oil received from household "do-it-yourself" generators) may be burned on site in a commercial space heater. Used oil also may be burned for energy recovery. Contact state and local authorities to determine requirements and to obtain necessary permits.

Oil filters   Drain for at least 24 hours, crush, and recycle used oil filters.

Batteries   Recycle batteries by sending them to a reclaimer or back to the distributor. Keeping shipping receipts can demonstrate that you have recycled. Store batteries in a watertight, acid-resistant container. Inspect batteries for cracks and leaks when they come in. Treat a dropped battery as if it were cracked. Acid residue is hazardous because it is corrosive and may contain lead and other toxics. Neutralize spilled acid by using baking soda or lime, and dispose of it as hazardous material.

Metal residue from machining   Collect metal filings when machining metal parts. Keep separate and recycle if possible. Prevent metal filings from falling into a storm sewer drain.

Solvents   Replace hazardous chemicals with less toxic alternatives that have equal performance. For example, substitute water-based cleaning solvents for petroleum-based solvent degreasers. To reduce the amount of solvent used when cleaning parts, use a two-stage process (dirty solvent followed by fresh solvent). Hire a hazardous waste management service to clean and recycle solvents. (Some spent solvents must be disposed of as hazardous waste unless recycled properly). Store solvents in closed containers to prevent evaporation. Evaporation of solvents contributes to ozone depletion and smog formation. In addition, the residue from evaporation must be treated as a hazardous waste. Properly label spent solvents and store in drip pans or in diked areas and only with compatible materials.

Other solids   Store materials such as scrap metal, old machine parts, and worn tires under a roof or tarpaulin to protect them from the elements and to prevent potentially contaminated runoff. Consider recycling tires by retreading them.

Liquid recycling   Collect and recycle coolants from radiators. Store transmission fluids, brake fluids, and solvents containing chlorinated hydrocarbons separately, and recycle or dispose of them properly.

Shop towels and rags   Keep waste towels in a closed container marked "Contaminated shop towels only." To reduce costs and liabilities associated with disposal of used towels, which can be classified as hazardous wastes, investigate using a laundry service that is able to treat the wastewater generated from cleaning the towels.

Waste storage   Always keep hazardous waste separate, properly labeled, and sealed in the recommended containers. The storage area should be covered and may need to be fenced and locked if vandalism could be a problem. Select a licensed hazardous waste hauler after seeking recommendations and reviewing the firm's permits and authorizations.

NATEF TASK LIST FOR HEATING AND AIR CONDITIONING

The priority rating indicates the minimum percentage of tasks, by area, a program must include in its curriculum in order to be certified in that area. Priority rating 1 has the most importance.

A. A/C System Diagnosis and Repair
   A.1. Identify and interpret heating and air conditioning concerns; determine necessary action. Priority Rating 1.
   A.2. Research applicable vehicle and service information, such as heating and air conditioning operation, vehicle service history, service precautions, and technical service bulletins. Priority Rating 1.
   A.3. Locate and interpret vehicle and major component identification numbers (VIN, vehicle certification labels, calibration labels). Priority Rating 1.
   A.4. Performance test A/C system; diagnose A/C system malfunctions using principles of refrigeration. Priority Rating 1.
   A.5. Diagnose abnormal operating noises in the A/C system; determine necessary action. Priority Rating 2.
   A.6. Identify refrigerant type; conduct a performance test of the A/C system; determine necessary action. Priority Rating 1.
   A.7. Leak test A/C system; determine necessary action. Priority Rating 1.
   A.8. Inspect the condition of discharged oil; determine necessary action. Priority Rating 2.
   A.9. Determine recommended oil for system application. Priority Rating 2.
B. Refrigeration System Component Diagnosis and Repair
   1.   Compressor and Clutch
      B.1.1. Diagnose A/C system conditions that cause the protection devices (pressure, thermal, and PCM) to interrupt system operation; determine necessary action. Priority Rating 2.
      B.1.2. Inspect A/C compressor drive belts; determine necessary action. Priority Rating 2.
      B.1.3. Inspect, test, and/or replace A/C compressor clutch components and/or assembly. Priority Rating 2.
      B.1.4. Remove and reinstall A/C compressor and mountings; measure oil quantity; determine necessary action. Priority Rating 2.

2.   *Evaporator, Condenser, and Related Components*
    B.2.1. Determine need for an additional A/C system filter; perform necessary action. Priority Rating 1.
    B.2.2. Remove and inspect AC system mufflers, hoses, lines, fittings, O-rings, seals, and service valves; perform necessary action. Priority Rating 2.
    B.2.3. Inspect A/C condenser for airflow restrictions; perform necessary action. Priority Rating 1.
    B.2.4. Remove and install receiver/drier or accumulator/drier; measure oil quantity; determine necessary action. Priority Rating 2.
    B.2.5. Remove and install expansion valve or orifice (expansion) tube. Priority Rating 2.
    B.2.6. Inspect evaporator housing water drain; perform necessary action. Priority Rating 3.
    B.2.7. Remove and reinstall evaporator; measure oil quantity; determine necessary action. Priority Rating 1.
    B.2.8. Remove and reinstall condenser; measure oil quantity; determine necessary action. Priority Rating 1.

C. Heating, Ventilation, and Engine Cooling Systems Diagnosis and Repair
    C.1. Diagnose temperature control problems in the heater/ventilation system; determine necessary action. Priority Rating 2.
    C.2. Perform cooling system, cap, and recovery system tests (pressure, combustion leakage, and temperature); determine necessary action. Priority Rating 1.
    C.3. Inspect engine cooling and heater system hoses and belts; perform necessary action. Priority Rating 1.
    C.4. Inspect, test, and replace thermostat and housing. Priority Rating 1.
    C.5. Determine coolant condition and coolant type for vehicle application; drain and recover coolant. Priority Rating 1.
    C.6. Flush system; refill system with recommended coolant; bleed system. Priority Rating 1.
    C.7. Inspect and test cooling fan, fan clutch (electrical and mechanical), fan shroud, and air dams; perform necessary action. Priority Rating 1.
    C.8. Inspect and test electrical cooling fan, fan control system, and circuits; determine necessary action. Priority Rating 1.
    C.9. Inspect and test heater control valve(s); perform necessary action. Priority Rating 2.
    C.10. Remove and reinstall heater core. Priority Rating 1.

D. Operating Systems and Related Controls Diagnosis and Repair
    D.1. Diagnose malfunctions in the electrical controls of heating, ventilation, and A/C (HVAC) systems; determine necessary action. Priority Rating 2.
    D.2. Inspect and test A/C-heater blower, motors, resistors, switches, relays, wiring, and protection devices; perform necessary action. Priority Rating 2.
    D.3. Test and diagnose A/C compressor clutch control systems; determine necessary action. Priority Rating 3.
    D.4. Diagnose malfunctions in the vacuum and mechanical components and controls of the HVAC system; determine necessary action. Priority Rating 2.
    D.5. Inspect and test A/C-heater control panel assembly; determine necessary action. Priority Rating 3.
    D.6. Inspect and test A/C-heater control cables and linkages; perform necessary action. Priority Rating 3.
    D.7. Inspect and test A/C-heater ducts, doors, hoses, cabin filters, and outlets; perform necessary action. Priority Rating 3.
    D.8. Check operation of automatic and semiautomatic HVAC control systems; determine necessary action. Priority Rating 3.

E. Refrigerant Recovery, Recycling, and Handling
    E.1. Verify correct use and maintenance of refrigerant handling equipment. Priority Rating 1.
    E.2. Identify (by label application or use of a refrigerant identifier) and recover A/C system refrigerant. Priority Rating 1.
    E.3. Recycle refrigerant. Priority Rating 1.
    E.4. Label and store refrigerant. Priority Rating 1.

E.5.   Test recycled refrigerant for noncondensable gases. Priority Rating 1.

E.6.   Evacuate and charge A/C system. Priority Rating 1.

DEFINITION OF TERMS USED IN THE TASK LIST

To clarify the intent of these tasks, NATEF has defined some of the terms used in the task listings. For a good understanding of what the task includes, refer to this glossary while reading the task list.

add	To increase fluid or pressure to the correct level or amount.
adjust	To bring components to specified operational settings.
assemble (reassemble)	To fit together the components of a device.
bleed	To remove air from a closed system.
charge	To bring to "full" state (e.g., battery or air conditioning system).
check	To verify condition by performing an operational or comparative examination.
clean	To rid components of extraneous matter for the purpose of reconditioning, repairing, measuring, and reassembling.
determine	To establish the procedure to be used to effect the necessary repair.
determine necessary action	Indicates that the diagnostic routine or routines is the primary emphasis of a task. The student is required to perform the diagnostic steps and communicate the diagnostic outcomes and corrective actions required to address the concern or problem. The training program determines the communication method (worksheet, test, verbal communication, or other means deemed appropriate) and whether the corrective procedures for these tasks are actually performed.
diagnose	To locate the root cause or nature of a problem by using the specified procedure.
disassemble	To separate a component's parts in preparation for cleaning, inspection, or service.
discharge	To empty a storage device or system.
drain	To use gravity to empty a container.
evacuate	To remove air, fluid, or vapor from a closed system by the use of a vacuum pump.
fill (refill)	To bring fluid level to a specified point or volume.
flush	To use fluid to clean an internal system.
identify	To establish the identity of a vehicle or component prior to service; to determine the nature or degree of a problem.
inspect	(see *check*)
install (reinstall)	To place a component in its proper position in a system.
leak test	To locate the source of leaks in a component or system.
perform	To accomplish a procedure in accordance with established methods and standards.
perform necessary action	Indicates that the student is to perform the diagnostic routine(s) and perform the corrective action item. Where various scenarios (conditions or situations) are presented in a single task, at least one of the scenarios must be accomplished.
pressure test	To use air or fluid pressure to determine the integrity, condition, or operation of a component or system.

purge	To eliminate an undesired air or fluid from a closed system.
reassemble	(See *assemble*)
refill	(See *fill*)
remove	To disconnect and separate a component from a system.
replace	To exchange an unserviceable component with a new or rebuilt component; to reinstall a component.
select	To choose the correct part or setting during assembly or adjustment.
service	To perform a specified procedure when called for in the owner's manual or service manual.
test	To verify condition through the use of meters, gauges, or instruments.
verify	To establish that a problem exists after hearing the customer's complaint and performing a preliminary diagnosis.

HEATING AND AIR CONDITIONING TOOLS AND EQUIPMENT

Many different tools and pieces of testing and measuring equipment are used to service heating and air conditioning systems. NATEF has said that an air conditioning technician must know what they are and how and when to use them. The tools and equipment listed by NATEF are covered in the following discussion, along with the tools and equipment you will use while completing the job sheets. Although you will be using common hand tools, they are not part of this discussion. You should already know what they are and how to use and care for them.

NOTE: *Air conditioning systems are extremely sensitive to moisture and dirt. Therefore, clean working conditions are extremely important. The smallest particle of foreign matter in an air conditioning system contaminates the refrigerant, causing rust, ice, or damage to the compressor. For this reason, all replacement parts are sold in vacuum-sealed containers and should not be opened until they are to be installed in the system. If, for any reason, a part has been removed from its container for any length of time, the part must be completely flushed using only a recommended flush solvent to remove any dust or moisture that might have accumulated during storage. When the system has been open for any length of time, the entire system must be purged completely. A new receiver/drier must be installed because the element of the existing unit will have become saturated and unable to remove any moisture from the system once the system is recharged.*

REFRIGERANT SAFETY PRECAUTIONS

- Always work in a well-ventilated and clean area. Refrigerant (R-12 and R-134a) is colorless and invisible as a gas. Refrigerant is heavier than oxygen and will displace oxygen in a confined area. Avoid breathing the refrigerant vapors. Exposure to refrigerant may irritate your eyes, nose, and throat.

- Refrigerant evaporates quickly when it is exposed to the atmosphere. It will freeze anything it contacts. If liquid refrigerant gets in your eyes or on your skin, it can cause frostbite. Never rub your eyes or skin if you have come in contact with refrigerant. Immediately flush the exposed areas with cool water for 15 minutes and seek medical help.

- An A/C system's high pressure can cause severe injury to your eyes or skin if a hose bursts. Always wear eye protection when working around the A/C system and refrigerant. It is also advisable to wear protective gloves and clothing.

- Never use R-134a in combination with compressed air for leak testing. Pressurized R-134a, in the presence of oxygen, may form a combustible mixture. Never introduce compressed air into R-134a containers (empty or full ones), A/C systems, or A/C service equipment.

- Be careful when handling refrigerant containers. Never drop, strike, puncture, or burn the containers. Always use Department of Transportation (DOT)-approved refrigerant containers.

■ Never expose A/C system components to high temperatures—heat causes the refrigerant's pressure to increase. Never expose refrigerant to an open flame—a poisonous gas (phosgene) results. If inhaled, phosgene gas can cause severe respiratory irritation. This gas also occurs when an open-flame leak detector is used. Phosgene fumes have an acrid (bitter) smell. If the refrigerant needs to be heated during service, the bottom of the refrigerant container should be placed in warm water (less than 125°F).

■ Never overfill refrigerant containers. The filling level of the container should never exceed 60 percent of the container's gross weight rating. Always store refrigerant containers in temperatures below 125°F and keep them out of direct sunlight.

■ Refrigerant comes in 30- and 50-pound cylinders. Remember that these drums are under considerable pressure and should be handled following precautions. Keep the drums in an upright position. Make sure that metal caps protect valves and safety plugs when the drums are not in use. Avoid dropping the drums. Handle them carefully. When transporting refrigerant, do not place containers in the vehicle's passenger compartment.

■ R-12 should be stored and sold in white containers, whereas R-134a should be stored in light blue containers. R-12 and R-134a should never be mixed. Their oils and desiccants are not compatible. If the two refrigerants are mixed, contamination will occur and may result in A/C system failure. Separate service equipment, including recovery/recycling machines and service gauges, should be used for the different refrigerants. Always read and follow the instructions from the equipment manufacturer when servicing A/C systems.

■ To prevent cross-contamination, identify whether the A/C system being worked on uses R-12 or R-134a. Check the fittings in the system; all R-134a-based systems use 1/2-inch-16 ACME threaded fittings and quick-disconnect service couplings. Underhood labels clearly stating that R-134a is used can identify most R-134a systems. Most manufacturers identify the type of refrigerant used by labeling the compressor. Also look for a label with the words, *"CAUTION—SYSTEM TO BE SERVICED BY QUALIFIED PERSONNEL."* This label or plate can be found under the hood, near a component of the system. This label also tells you what kind of refrigerant and refrigerant oil are used. Before beginning to service an A/C system, determine the type of refrigerant used.

■ R-134a can also be identified by viewing the sight glass, if the system has one. The appearance of R-134a will be milky due to the mixture of refrigerant and the refrigerant oil.

Manifold Gauge Set

The manifold gauge set (Figure 2) is used when discharging, charging, evacuating, and diagnosing trouble in the system. With the new legislation on handling refrigerants, all gauge sets are required to have a valve device to close off the end of the hose so that the fitting not in use is automatically shut.

The low-pressure gauge is graduated into pounds of pressure from 1 to 120 (with cushion to 250) in 1-pound graduations, and, in the opposite direction, in inches of vacuum from 0 to 30. This is the gauge that should always be used in checking pressure on the low-pressure side of the system. The high-pressure gauge is graduated from 0 to 500 pounds pressure in 10-pound graduations. This gauge is used for checking pressure on the high-pressure side of the system.

The center manifold fitting is common to both the low and the high side and is for evacuating or adding refrigerant to the system. When this fitting is not being used, it should be capped. A test hose connected to the fitting directly under the low-side gauge is used to connect the low side of the test manifold to the low side of the system. A similar connection is found on the high side.

The gauge manifold is designed to control refrigerant flow. When the manifold test set is connected into the system, pressure is registered on both gauges at all times. During all tests, both the low- and high-side hand valves are in the closed position (turned inward until the valve is seated).

Refrigerant flows around the valve stem to the respective gauges and registers the system's low-side pressure on the low-side gauge and the system's high-side pressure on the high-side gauge. The hand valves isolate the low and high sides from the central portion of the manifold. When the gauges are first connected to the gauge fittings with the refrigeration system charged, the gauge lines should always be

Figure 2 A manifold gauge set.

purged. Purging is done by cracking each valve on the gauge set to allow the pressure of the refrigerant in the refrigeration system to force the air to escape through the center gauge line. Failure to purge lines can result in air or other contaminants entering the refrigeration system.

Because R-134a is not interchangeable with R-12, separate sets of hoses, gauges, and other equipment are required to service vehicles. All equipment used to service R-134a and R-12 systems must meet Society of Automotive Engineers (SAE) standard J1991. The service hoses on the manifold gauge set must have manual or automatic back-flow valves at the service port connector ends. These valves prevent the refrigerant from being released into the atmosphere during connection and disconnection. Manifold gauge sets for R-134a can be identified by labels on the gauges; in addition or instead, the face of the gauges may be light blue.

For identification purposes, R-134a service hoses must have a black stripe along their length and be clearly labeled. The low-pressure hose is blue with a black stripe. The high-pressure hose is red with a black stripe, and the center service hose is yellow with a black stripe. Service hoses for one type of refrigerant do not easily connect to the wrong system because the fittings for an R-134a system are different from those used in an R-12 system.

Service Port Adapters

The high-side fitting on many vehicles with an R-12 system may require the use of a special adapter to connect the manifold gauge set to the service port. These adapters (Figure 3) are installed to the end

Figure 3 High-side adapters to connect the service hose to the service ports.

of the service hose before connecting the hose to the system. The service hoses of some manifold gauge sets are not equipped with a Schrader valve-depressing pin. Therefore, when connecting this type of hose to a Schrader valve, an adapter (Figure 4) must be used. The manifold and gauge sets for R-12 and R-134a are not interchangeable, therefore there are no suitable or allowable adapters for using R-12 gauges on a R-134a system, or vice versa.

Electronic Leak Detector

An electronic leak detector is safe, effective, and can be used with all types of refrigerants. A hand-held, battery-operated electronic leak detector contains a test probe that is moved about 1 inch per second in areas of suspected leaks (Figure 5). Since refrigerant is heavier than air, the probe should be positioned

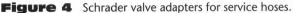

Figure 4 Schrader valve adapters for service hoses.

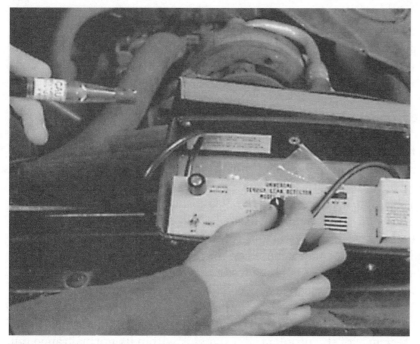

Figure 5 An electronic leak detector.

below the test point. An alarm or a buzzer on the detector indicates the presence of a leak. On some models, a light flashes when refrigerant is detected.

Fluorescent Leak Tracer

To find a refrigerant leak using the fluorescent tracer system, first introduce a fluorescent dye into the air conditioning system with a special infuser included with the detector equipment. Run the air conditioner for a few minutes, giving the tracer dye fluid time to circulate and penetrate. Wearing the tracer protective goggles, scan the system with a black-light glow gun. Leaks in the system shine under the black light as a luminous yellow-green.

Scan Tools

The introduction of computer-controlled systems brought with it the need for tools capable of troubleshooting electronic control systems. A variety of computer scan tools are available today that do just that. A scan tool is a microprocessor designed to communicate with the vehicle's computer. Connected to the computer through diagnostic connectors, a scan tool can access trouble codes, run tests to check system operations, and monitor the activity of the system. Trouble codes and test results are displayed on an LED screen or printed out on the scanner printer.

Scan tools retrieve fault codes from a computer's memory and digitally display these codes on the tool. A scan tool may also perform many other diagnostic functions depending on the year and make of the vehicle. Most aftermarket scan tools have removable modules that are updated each year. These modules are designed to test the computer systems on various makes of vehicles. For example, some scan testers have a 3-in-1 module that tests the computer systems on Chrysler, Ford, and General Motors vehicles. A 10-in-1 module is also available to diagnose computer systems on vehicles imported by ten different manufacturers. These modules plug into the scan tool.

Scan tools are capable of testing many onboard computer systems, such as climate controls, transmission controls, engine computers, antilock brake computers, air bag computers, and suspension computers, depending on the year and make of the vehicle and the type of scan tester. In many cases, the technician must select the computer system to be tested with the scanner after it has been connected to the vehicle.

The scan tool is connected to specific diagnostic connectors on various vehicles. Most manufacturers have one diagnostic connector that connects the data wire from each onboard computer to a specific terminal in this connector. Other vehicle manufacturers have several diagnostic connectors on each vehicle, and each of these connectors may be connected to one or more onboard computers. A set of connectors is supplied with the scanner to allow tester connection to various diagnostic connectors on different vehicles.

The scanner must be programmed for the model year, make of vehicle, and type of engine. With some scan tools, this selection is made by pressing the appropriate buttons on the tester, as directed by the digital tester display. On other scan testers, the appropriate memory card must be installed in the tester for the vehicle being tested. Some scan testers have a built-in printer to print test results, whereas other scan testers may be connected to an external printer.

As automotive computer systems become more complex, the diagnostic capabilities of scan testers continue to expand. Many scan testers now have the capability to store, or *freeze,* data into the tester during a road test, and then play the data back when the vehicle is returned to the shop.

Some scan testers now display diagnostic information based on the fault code in the computer memory. Service bulletins published by the manufacturer of the scan tester may be indexed by the tester after the vehicle information is entered in the tester. Other scan testers will display sensor specifications for the vehicle being tested.

The vehicle's computer sets trouble codes when a voltage signal is entirely out of its normal range. The codes help technicians identify the cause of the problem. If a signal is within its normal range but is still not correct, the vehicle's computer does not display a trouble code, though a problem may still exist.

With the On-Board Diagnostic II system (OBD-II), the diagnostic connectors are located in the same place on all vehicles. Also, any scan tools designed for OBD-II work on all OBD-II systems; therefore, the need for designated scan tools or cartridges is eliminated. The OBD-II scan tool has the ability to run diagnostic tests on all systems and has "freeze-frame" capabilities.

Stethoscope

Some sounds can be easily heard without using a listening device, but others are impossible to hear unless amplified. A stethoscope is very helpful in locating the cause of a noise by amplifying the sound waves. It can also help you distinguish between normal and abnormal noise. The procedure for using a stethoscope is simple. Use the metal prod to trace the sound until it reaches its maximum intensity. Once the precise location has been discovered, the sound can be better evaluated. A sounding stick, which is nothing more than a long, hollow tube, works on the same principle, though a stethoscope gives much clearer results.

The best results, however, are obtained with an electronic listening device. With this tool you can tune into the noise. Doing this allows you to eliminate all other noises that might distract or mislead you.

Belt Tension Gauge

A belt tension gauge is used to measure drive belt tension. The belt tension gauge is installed over the belt, and the gauge indicates the amount of belt tension.

Coolant Hydrometer

A coolant hydrometer is used to check the amount of antifreeze in the coolant. This tester contains a pickup hose, coolant reservoir, and squeeze bulb. The pickup hose is placed in the radiator coolant. When the squeeze bulb is squeezed and released, coolant is drawn into the reservoir. As coolant enters the reservoir, a pivoted float moves upward with the coolant level. A pointer on the float indicates the freezing point of the coolant on a scale located on the reservoir housing.

Cooling System Pressure Tester

A cooling system pressure tester contains a hand pump and a pressure gauge. A hose is connected from the hand pump to a special adapter that fits on the radiator filler neck. This tester is used to pressurize the cooling system and check for coolant leaks. Additional adapters are available to connect the tester to the radiator cap. With the tester connected to the radiator cap, the pressure-relief action of the cap may be checked.

Thermometer

A dial-type thermometer (Figure 6) is often used to measure the air temperature at the vent outlets. This procedure checks the overall performance of the system. The thermometer can also be used to check the temperature of refrigerant lines, hoses, and components while diagnosing a system. An electronic pyrometer works best for this procedure and is often used.

Feeler Gauge

A feeler gauge is a thin strip of metal or plastic of known and closely controlled thickness. Several of these strips are often assembled together as a feeler gauge set, which looks like a pocketknife. The desired thickness gauge can be pivoted away from others for convenient use. A feeler gauge set usually contains strips or leaves of 0.002- to 0.010-inch thickness (in steps of 0.001 inch) and leaves of 0.012- to 0.024-inch thickness (in steps of 0.002 inch).

A feeler gauge can be used by itself to measure clearances, such as critical compressor clutch clearances. It can also be used with a precision straightedge to measure the flatness of sealing surfaces in a compressor.

Straightedge

A straightedge is no more than a flat bar machined to be totally flat and straight; to be effective, it must be flat and straight. Any surface that should be flat can be checked with a straightedge and feeler gauge set. The straightedge is placed across and at angles on the surface. At any low points on the surface, a

Figure 6   Three types of commonly used thermometers.

feeler gauge can be placed between the straightedge and the surface. The size gauge that fills in the gap is the amount of warpage or distortion.

Dial Indicator

The dial indicator is calibrated in 0.001-inch (one-thousandth-inch) increments. Metric dial indicators are also available; both types are used to measure movement. Common uses of the dial indicator include checking compressor shaft run out, shaft endplay, and compressor clutch run out.

To use a dial indicator, position the indicator rod against the object to be measured. Then, push the indicator toward the work until the indicator needle travels far enough around the gauge face to permit movement to be read in either direction. Zero the indicator needle on the gauge. Move the object in the direction required while observing the needle of the gauge. Always be sure the range of the dial indicator is sufficient to allow the amount of movement required by the measuring procedure. For example, never use a 1-inch indicator on a component that will move 2 inches.

Torque-Indicating Wrench

Torque is the twisting force used to turn a fastener against the friction between the threads and between the head of the fastener and the surface of the component. The fact that practically every vehicle and engine manufacturer publishes a list of torque recommendations is ample proof of the importance of using proper amounts of torque when tightening nuts or bolts. The amount of torque applied to a fastener is measured with a torque-indicating, or torque, wrench.

Three basic types of torque-indicating wrenches are available, in pound-per-inch and pound-per-foot increments. A beam torque wrench has a beam that points to the torque reading; a click-type torque wrench has the desired torque reading set on the handle (when the torque reaches that level, the wrench clicks); and a dial torque wrench has a dial that indicates the torque exerted on the wrench. Some designs of the dial torque wrench have a light or buzzer that turns on when the desired torque is reached.

Gear and Bearing Pullers

Many tools are designed for a specific purpose. An example of a special tool is a gear and bearing puller. Many gears and bearings have a slight interference fit (press fit) when they are installed on a shaft or in a housing. Something that has a press fit has an interference fit. For example, the inside diameter of a bore is 0.001 inch smaller than the outside diameter of a shaft, and when the shaft is fitted into the bore, it must be pressed in to overcome the 0.001-inch interference fit. This press fit prevents the parts from moving on each other. The removal of these gears and bearings must be done carefully to prevent damage to the gears, bearings, or shafts. Prying or hammering can break or bind the parts. A puller with the proper jaws and adapters should be used to remove gears and bearings. Using the proper puller, the force required to remove a gear or bearing can be applied with a slight and steady motion.

Bushing and Seal Pullers and Drivers

Another commonly used group of special tools are the various designs of bushing and seal drivers and pullers. Pullers are either a threaded or slide-hammer type of tool. Always make sure you use the correct tool for the job because bushings and seals are easily damaged if the wrong tool or procedure is used. Car manufacturers and specialty tool companies work together closely to design and manufacture special tools required to repair cars. Most of these special tools are listed in the appropriate service manuals.

Retaining Ring Pliers

An air conditioning technician will encounter many different styles and sizes of retaining rings and snap rings that hold things together or keep them in a fixed location. Using the correct tool to remove and install these rings is the only safe way to work with them. All technicians should have an assortment of snap ring and retaining ring pliers.

Compressor Tools

Although compressors are typically replaced when they are faulty, certain service procedures applied to them are standard practice. Most of these procedures focus on compressor clutch and shaft seal service, and they require special tools (Figure 7). Clutch plate tools are required to gain access to the shaft seal. They are also needed to reinstall the clutch plate after service.

Many different tools are required to perform these services on a compressor. Typically, to replace a shaft seal, you need an adjustable or fixed spanner wrench, clutch plate installer/remover, ceramic seal installer/remover, seal assembly installer/remover, seal seat installer/remover, shaft seal protector, snap ring pliers, O-ring remover, and O-ring installer. Some of these tools are made for a specific model compressor; others are universal fit or will have interchangeable parts to allow them to work on a variety of compressors.

Can Tap

A can tap is used to connect a 1-"pound" can of refrigerant to the manifold gauge set. These cans are available in either screw top or flat top. The can tap either pierces through the top or makes a screw-type connection with the top. It should be noted that "1-pound" cans actually contain only 12 or 14 ounces. Many states have restrictions or strict prohibitions on the sale and use of small 1-pound cans of refrigerant.

Hose and Fitting Tools

An A/C system is a closed system, meaning outside air should never enter the system, and the refrigerant in the system should never exit to the outside. To maintain its integrity, special fittings and hoses are used. Often, special tools are required when servicing the system's fittings and hoses. An example of this is the spring-lock coupling tool set; without the tool, it is impossible to separate this connector without damaging it.

Figure 7 A sample of common compressor tools.

Charging Cylinder

The charging cylinder is designed to mete out a desired amount of refrigerant by weight. Compensation for temperature variations is accomplished by reading the pressure on the gauge of the cylinder and dialing the plastic shroud. The calibrated chart on the shroud contains corresponding pressure readings for the refrigerant being used.

Refrigerant Identifier

A refrigerant identifier (Figure 8), obviously, is used to identify the type of refrigerant that is present in a system. This needs to be done before any service work begins. The tester is used to identify the purity and quality of the refrigerant sample taken from the system.

Refrigerant Recovery/Recycling System

Each shop that does A/C work must have a recovery, recycle, and recharge machine (Figure 9) for each type of refrigerant. There are some recovery, recycle, and recharge machines that have the capability of working with R-12 and R-134a, but check the machine before attempting to do so.

There are currently two types of refrigerant recovery machines, the single-pass and the multipass. Both have the ability to draw the refrigerant from the vehicle, filter and separate the oil from it, remove moisture and air from it, and store the refrigerant until it is reused.

In a single-pass system, the refrigerant goes through each stage before being stored. In multipass systems, the refrigerant may go through all stages or some of the stages before being stored. Either system is acceptable if it has the UL-approved label.

All recycled refrigerant must be safely stored in DOT CFR Title 49-approved or UL-approved containers. Containers specifically made for R-134a should be so marked. Before any container of recycled refrigerant can be used, it must be checked for noncondensable gases.

Vacuum Pump

It is critical that all moisture be removed from a system before it is recharged. Moisture is removed by connecting a vacuum pump to the system. The vacuum pump may be a separate item or may be incorporated into other air conditioning equipment, such as the recovery/recycling machine.

Figure 8 A refrigerant identifier.

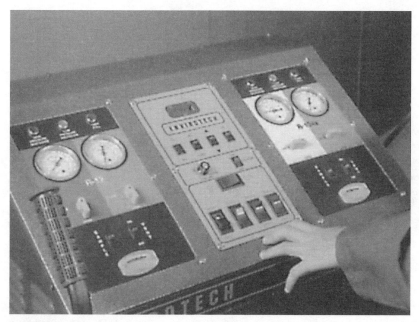

Figure 9 A refrigerant recovery/recycling machine.

The main purpose of the vacuum pump is to remove the contaminating air and moisture from the system. The vacuum pump reduces system pressure in order to vaporize the moisture and then exhausts the vapor along with all remaining air. The pump's ability to clean the system is directly related to its ability to reduce pressure (create a vacuum) low enough to boil off all the contaminating moisture.

Thermistor Vacuum Gauge

An electronic thermistor vacuum gauge is designed to work with the vacuum pump to measure the last, most critical inch of mercury during evacuation. It constantly monitors and visually indicates the vacuum level so that you know when the system has a full vacuum and will be free of moisture.

After the system is evacuated, it can be recharged. If the system will not pull down to a good vacuum, there is probably a leak somewhere in the system.

Electronic Scale

An electronic scale (Figure 10) is used to ensure that an accurate amount of refrigerant has been installed in a system. Since refrigerant is installed by weight, precise measurements of the refrigerant container before and after charging the system indicates how much refrigerant has been added to the system.

Coolant Recovery and Recycle System

A coolant recovery and recycle machine typically can drain, recycle, fill, flush, and pressure test a cooling system. Typically, additives are mixed into the used coolant during recycling. These additives either bind to contaminants in the coolant so they can be easily removed, or they restore some of the chemical properties in the coolant.

Service Manuals

Perhaps the most important tools you will use are service manuals. There is no way a technician can remember all the procedures and specifications needed to correctly repair all vehicles. Thus, a good technician relies on service manuals and other information sources for this information. Good information plus knowledge allows a technician to fix a problem with the least amount of frustration and at the lowest cost to the customer.

Figure 10   An electronic scale.

To obtain the correct system specifications and other information, you must first identify the exact system you are working on. The best source for vehicle identification is the vehicle identification number (VIN). The code can be interpreted through information given in the service manual. In addition, the manual may also help you recognize the system through identification of key components or other numbers or markings.

The primary source of repair and specification information for any car, van, or truck is the manufacturer. The manufacturer publishes service manuals each year for every vehicle built. Because of the enormous amount of information, some manufacturers publish more than one manual per year per car model. They are typically divided into sections based on the major systems of the vehicle. In the case of air conditioning systems, there is a section for each engine that may be in the vehicle. Often the heating system is included with information about the engine's cooling system. Manufacturers' manuals cover all repairs, adjustments, specifications, detailed diagnostic procedures, and the special tools required.

Since many technical changes occur on specific vehicles each year, manufacturers' service manuals must be constantly updated. Updates are published as service bulletins (often referred to as Technical Service Bulletins, or TSBs) that show the changes in specifications and repair procedures during the model year. These changes do not appear in the service manual until the next year. The car manufacturer provides these bulletins to dealers and repair facilities on a regular basis.

Service manuals are also published by independent companies rather than the manufacturers. However, they pay for and get most of their information from the car makers. They contain component information, diagnostic steps, repair procedures, and specifications for several car makes in one book. Information is usually condensed and is more general in nature than that in the manufacturer's manuals. The condensed format allows for more coverage in less space and, therefore, is not always specific. One book may cover several years of models and several car makes.

Many of the larger parts manufacturers have excellent guides on the various parts they manufacture or supply. They also provide updated service bulletins on their products. Other sources for up-to-date technical information are trade magazines and trade associations.

The same information that is available in service manuals and bulletins is also available on CD-ROMs and DVDs. A single compact disk can hold 250,000 pages of text. This eliminates the need for a huge library to contain all of the printed manuals. Using a CD-ROM to find information is also easier and quicker. The disks are normally updated monthly and contain the most recent service bulletins, along with engineering and field service fixes.

CROSS-REFERENCE GUIDE

NATEF Task	Job Sheet
A.1	28
A.2	29
A.3	29
A.4	30
A.5	1
A.6	2
A.7	3
A.8	4
A.9	4
B.1.1	5
B.1.2	1
B.1.3	5
B.1.4	6
B.2.1	7
B.2.2	8
B.2.3	1
B.2.4	8
B.2.5	8
B.2.6	9
B.2.7	31
B.2.8	32
C.1	10
C.2	11
C.3	12
C.4	13
C.5	11
C.6	11
C.7	14
C.8	15
C.9	16
C.10	33
D.1	17
D.2	18
D.3	19
D.4	17
D.5	20

HEATING AND AIR CONDITIONING JOB SHEET 2

Identifying the Type of Air-Conditioning System in a Vehicle

Name _____ Station _____ Date _____

NATEF Correlation

This Job Sheet addresses the following NATEF task:

A.6. Identify refrigerant type; conduct a performance test of the A/C system; determine necessary action.

Objective

Upon completion of this job sheet, you will be able to use a service manual and the labeling in the engine compartment to identify the type of air conditioning used in a vehicle and check the system's overall performance.

Tools and Equipment

Service manual

Protective Clothing

Goggles or safety glasses with side shields

Describe the vehicle being worked on:

Year _____ Make _____ Model _____

VIN _____ Engine type and size _____

PROCEDURE

1. Look at the instrument panel controls for the air-conditioning system.   ☐ Task completed

2. Describe the controls:

   _____

   _____

3. Look at all of the labeling in the engine compartment and locate any label   ☐ Task completed
   that addresses the air conditioning system.

4. Summarize the contents of the labeling:

   _____

   _____

5. Look carefully at the fittings used at the compressor. What types of fittings are they?

_____

_____

_____

_____

6. What refrigerant is this system designed to use? _____ What type of refrigerant oil must be used? _____

7. Locate the information about this vehicle's air conditioning system in the   ☐ Task completed
   service manual.

8. Record the service manual and the pages where the information was found.

_____

_____

_____

9. Read through the material given in the service manual and determine what controls the flow of refrigerant through the evaporator. Summarize your findings here:

_____

_____

_____

10. Give a complete description of this air conditioning system:

_____

_____

_____

_____

_____

Problems Encountered

_____

_____

_____

Instructor's Comments

_____

_____

_____

HEATING AND AIR CONDITIONING JOB SHEET 4

Working with Refrigerant Oils

Name _____ Station _____ Date _____

NATEF Correlation

This Job Sheet addresses the following NATEF tasks:

A.8. Inspect the condition of discharged oil; determine necessary action.

A.9. Determine recommended oil for system application.

Objective

Upon completion of this job sheet, you will be able to inspect the condition of discharged oil and select the correct type of oil for a system. You will also be able to measure and add oil to an A/C system.

Tools and Materials
Graduated container
Service manual

Protective Clothing
Goggles or safety glasses with side shields

Describe the vehicle being worked on:
Year _____ Make _____ Model _____

VIN _____ Engine type and size _____

PROCEDURE

1. The compressor oil level is checked only where there is evidence of a major loss of system oil that could be caused by a broken refrigerant hose, severe hose fitting leak, badly leaking compressor seal, or collision damage to the system's components. Is this the situation?

   _____

   _____

2. When replacing refrigerant oil, it is important to use the specific type and quantity of oil recommended by the compressor manufacturer. If there is a surplus of oil in the system, too much oil circulates with the refrigerant, causing the cooling capacity of the system to be reduced. Too little oil results in poor lubrication of the compressor. When there has been excessive leakage or it is necessary to replace a component of the refrigeration system, certain procedures must be followed to ensure that the total oil charge in the system is correct after leak repair or the new part is on the car. Is this the situation?

   _____

   _____

3. To maintain the original total oil charge, it is necessary to compensate for this by adding oil to the new replacement part. Because of the differences in compressor designs, be sure to follow the manufacturer's instructions when adding refrigerant oil to their unit. Refer to the service manual and give a summary of the amount of oil that should be added for each component that may be replaced.

_____

_____

_____

_____

4. Name the type of oil that should be used in this system. Normally, R-12 based systems use mineral oil, while R-134a systems use synthetic polyalkaline glycol (PAG) oils. Using a mineral oil with R-134a will result in A/C compressor failure because of poor lubrication. Use only the oil specified for the system.

_____

5. If a component is being replaced and you are unsure of how much should be added to the system with the new component, completely drain the oil from the old component and measure it in a graduated container. Add that amount to the component when installing it.

☐ Task completed

_____

_____

Instructor's Comments

_____

_____

_____

HEATING AND AIR CONDITIONING JOB SHEET 5

Inspect, Test, and Service a Compressor Clutch

Name _____ Station _____ Date _____

NATEF Correlation

This Job Sheet addresses the following NATEF tasks:

B.1.1. Diagnose A/C system conditions that cause the protection devices (pressure, thermal, and PCM) to interrupt system operation; determine necessary action.

B.1.3. Inspect, test, and/or replace A/C compressor clutch components and/or assembly.

Objective

Upon completion of this job sheet, you will be able to inspect, test, and service a compressor clutch.

Tools and Materials

Fender covers

Hand tools

Necessary pullers and installers

Test light

DMM

Service manual

Protective Clothing

Goggles or safety glasses with side shields

Describe the vehicle being worked on:

Year _____ Make _____ Model _____

VIN _____ Engine type and size _____

Describe the type of air conditioning found on the vehicle:

_____

PROCEDURE

1. Place fender covers over the front fenders for protection.   ☐ Task completed

2. Visually inspect the clutch for damage.   ☐ Task completed

3. Check the electrical connections to the clutch.   ☐ Task completed

4. Using a test light or DMM with the AC system on, check the electrical plug   ☐ Task completed
   at the clutch for power.

5. If there is no power, locate and repair the problem. Retest.   ☐ Task completed

6. When power is present, check the ground circuit with a DMM.   ☐ Task completed

7. If it is determined that both power and ground are present at the compressor clutch, the clutch is defective and must be replaced.   ☐ Task completed

8. On some vehicles, the clutch can be serviced on the vehicle. On others, the compressor assembly must be removed to service the clutch. Refer to the service manual on your specific vehicle to determine what must be done.   ☐ Task completed

9. Using the proper pullers for your specific application and following the procedure in the service manual, remove the clutch.   ☐ Task completed

10. Reverse the procedure and install the new clutch.   ☐ Task completed

11. Test clutch operation after repairs are completed.   ☐ Task completed

Problems Encountered

_____

_____

_____

Instructor's Comments

_____

_____

_____

HEATING AND AIR CONDITIONING JOB SHEET 7

In-Line Filter Installation

Name _____ Station _____ Date _____

NATEF Correlation

This Job Sheet addresses the following NATEF task:

B.2.1. Determine need for an additional A/C system filter; perform necessary action.

Objective

Upon completion of this job sheet, you will be able to determine whether or not the A/C system needs a new filter and be able to correctly install one if necessary.

Tools and Materials
In-line filter assembly
Line wrenches
Service manual

Protective Clothing
Goggles or safety glasses with side shields

Describe the vehicle being worked on:
Year _____ Make _____ Model _____

VIN _____ Engine type and size _____

PROCEDURE

To determine if the system needs an in-line filter, answer the following questions:

1. Did the compressor need to be replaced because it was seized? _____

2. Does the system have a history of plugged metering devices? _____

3. If the answer to either of these questions was yes, the system needs to have a filter installed. Why would these conditions warrant a supplemental filter?

_____

_____

_____

4. Describe the in-line filter: does it have an orifice or is it just a filter?

_____

_____

_____

5. If the filter does not have an orifice, locate a convenient location for installation in the liquid line between the condenser outlet and the evaporator inlet. Describe where you will install the filter.

_____

_____

_____

6. If the filter has an orifice, find a convenient location for installation in the low-pressure side of the system beyond the expansion tube. Describe where you will install the filter.

_____

_____

_____

7. Make sure the system is void of all refrigerant. How did you do this?

_____

_____

_____

8. Secure the necessary tools, and then prepare the filter for installation according to the instructions that accompany the filter. What did you need to do?

_____

_____

_____

9. Install the filter into the appropriate line.   ☐ Task completed

10. If the filter had a built-in orifice, remove the original expansion tube from   ☐ Task completed
the system.

11. Make sure the lines are properly sealed, and then prepare to charge the system. What did you need to do before charging the system?

_____

_____

_____

Instructor's Comments

_____

_____

_____

HEATING AND AIR CONDITIONING JOB SHEET 9

Servicing the Evaporator Water Drain

Name _____ Station _____ Date _____

NATEF Correlation

This Job Sheet addresses the following NATEF task:

B.2.6. Inspect evaporator housing water drain; perform necessary action.

Objective

Upon completion of this job sheet, you will be able to inspect the evaporator housing water drain.

Tools and Materials

Drain pan
Clean rags
Air nozzle

Protective Clothing

Goggles or safety glasses with side shields

Describe the vehicle being worked on:

Year _____ Make _____ Model _____

VIN _____ Engine type and size _____

PROCEDURE

1. Raise the vehicle on a lift. ☐ Task completed

2. Locate the evaporator case drain tube and describe its overall condition.

   _____

   _____

3. Place the drain pan under the evaporator case. ☐ Task completed

4. Disconnect the tube from the case. Did water come out of the case?

   _____

5. If no water came out, carefully clean out the drain hole and tube at the ☐ Task completed
   evaporator case. If it is necessary to insert a rod through the tube to clean
   it out, do this carefully because the tube is accordion shaped and one of its
   bends may feel like a restriction.

6. If no water came out when you attempted to clear this tube, the entire evaporator case must be cleared. Describe your results so far.

_____

_____

7. If water came out of the case, the outside drain tube should be cleared by inserting the air nozzle into an end of the tube. Low air pressure should remove any restrictions in the tube. Describe what happened.

_____

_____

Instructor's Comments

_____

_____

_____

HEATING AND AIR CONDITIONING JOB SHEET 11

Test Cooling System

Name _____ Station _____ Date _____

NATEF Correlation

This Job Sheet addresses the following NATEF tasks:

C.2. Perform cooling system, cap, and recovery system tests (pressure, combustion leakage, and temperature); determine necessary action.

C.5. Determine coolant condition and coolant type for vehicle application; drain and recover coolant.

C.6. Flush system; refill system with recommended coolant; bleed system.

Objective

Upon completion of this job sheet, you will be able to perform tests on the cooling system, cap, and recovery system pressure; combustion leakage; and temperature, as well as test, drain, and recover coolant. You will also be able to flush and refill the cooling system with the recommended coolant and bleed the air as required.

Tools and Materials

Clean cloth rags
Cooling system tester
Coolant hydrometer
Drain pan
Hand tools
Cooling system flusher
Coolant recycler

Protective Clothing

Goggles or safety glasses with side shields

Describe the vehicle being worked on:

Year _____ Make _____ Model _____

VIN _____ Engine type and size _____

Describe general condition:

_____

PROCEDURE

1. Wipe off the radiator filler neck and inspect it.   ☐ Task completed

   Is the sealing seat free of accumulated dirt, nicks, or anything that might   ☐ Yes ☐ No
   prevent a good seal?

Are the cams on the out-turned flange bent or worn?

☐ Yes ☐ No
☐ Not applicable

2. Inspect the tube from the radiator to the expansion tank for dents and other obstructions. Run wire through the tube to be certain it is clear.

☐ Task completed

3. To test the cooling system for external leaks, attach a cooling system tester to the appropriate flexible adapter and carefully pump up pressure. Looking at tester's gauge, bring the pressure up to the proper cooling system test point indicated on the dial face.

☐ Task completed

Does the pressure begin to drop?

☐ Yes ☐ No

If so, check all external connections, including hoses, gaskets, and heater core for leaks.

☐ Task completed
☐ Not applicable

4. Check all points closely for small pinhole leaks or potentially dangerous weak points in the system.

☐ Task completed

To test for internal leaks:

5. Remove the flexible adapter from radiator, and replace the pressure cap.

☐ Task completed

6. Start the engine and allow it to reach normal operating temperature so the thermostat will open fully.

☐ Task completed

7. Slowly and carefully remove the pressure cap and replace it with the flexible adapter.

☐ Task completed

WARNING: *Engine coolant will be under pressure. Always wear hand protection and safety goggles or glasses with side shields when doing this task.*

8. Lock the tester slowly into place. Watch for pressure buildup.

☐ Task completed

WARNING: *If the pressure builds up suddenly due to an internal leak, remove the tester immediately.*

☐ Yes ☐ No

9. If no immediate pressure buildup is visible on the gauge, keep the engine running. Pressurize the cooling system.

☐ Task completed
☐ Not applicable

10. Does the gauge fluctuate? If so, the cooling system has a compression or combustion leak.

☐ Yes ☐ No

11. Before removing it, hold the tester body and press the pressure release button against some object on the car to relieve the pressure in the system.

☐ Task completed
☐ Not applicable

WARNING: *Always wear hand protection and safety goggles or glasses with side shields when doing this task. Avoid the hot coolant and steam.*

12. After the pressure has been relieved, shield your hands with a cloth wrapped around the flexible adapter and filler neck. Slowly turn the adapter cap from the lock position to the safety unlock position of the radiator filler neck cam. Do not detach the flexible adapter from the filler neck. Allow the pressure to dissipate completely in this safety position. Then you can remove the adapter and reinstall the radiator cap.

☐ Task completed
☐ Not applicable

13. Summarize your findings from these checks.

14. Draw some coolant out of the radiator or recovery tank with the coolant   ☐ Task completed
    hydrometer.

15. Check the strength of the coolant by observing the position of the tester's
    float. Summarize what you found out about the coolant and what service
    you recommend.

    _____

    _____

16. Reinstall the radiator cap, then start the engine and move the heater con-   ☐ Task completed
    trol to its full heat position.

17. Turn off the engine after it has only slightly warmed up.   ☐ Task completed

18. Place a drain pan under the radiator drain plug.   ☐ Task completed

19. Make sure the engine and cooling system is not hot, and then open the drain   ☐ Task completed
    plug. Keep the radiator cap on until the recovery tank is emptied.

20. Once all coolant has drained, close the drain plug. Then empty the drain   ☐ Task completed
    pan with the coolant into the coolant recycler and process the used coolant
    according to the manufacturer's instructions.

21. Connect the flushing machine to the cooling system.   ☐ Task completed

22. Follow the manufacturer's instructions for machine operation. Typically,   ☐ Task completed
    flushing continues until clear water flows out of the system. Once flushing
    is complete, disconnect the flushing machine.

23. What type of coolant is recommended for this vehicle? Where did you find
    the recommendation?

    _____

    _____

24. Look up the capacity of the cooling system in the service manual.

    What is it? _____

25. Prepare to add a mixture of 50% coolant and 50% water.   ☐ Task completed

26. Locate the engine's cooling system bleed valve. Where is it?

    _____

27. Loosen the bleed valve.   ☐ Task completed

28. Pour an amount of coolant that is equal to half of the cooling system's   ☐ Task completed
    capacity into the system.

29. Add water until some coolant begins to leak from the bleed valve. Close   ☐ Task completed
    the valve.

30. Leave the radiator cap off and start the engine.   ☐ Task completed

31. Continue to add water to the system as the engine warms up.   ☐ Task completed

32. Once the system appears full, install the radiator cap. ☐ Task completed

33. Observe the system for leaks and watch the activity in the recovery tank. ☐ Task completed
If no coolant moves to the tank, the system may need to be bled.

Problems Encountered

_____

_____

_____

Instructor's Comments

_____

_____

_____

HEATING AND AIR CONDITIONING JOB SHEET 12

Servicing Cooling System Hoses

Name _____ Station _____ Date _____

NATEF Correlation

This Job Sheet addresses the following NATEF task:

    C.3. Inspect engine cooling and heater system hoses and belts; perform necessary action.

Objective

Upon completion of this job sheet, you will be able to inspect and replace cooling and heater system hoses.

Tools and Materials

Basic hand tools
Knife
Wire brush
Emery cloth

Protective Clothing

Goggles or safety glasses with side shields

Describe the vehicle being worked on:

Year _____ Make _____ Model _____

VIN _____ Engine type and size _____

PROCEDURE

1. Carefully check all cooling hoses for leakage, swelling, and chafing. Record your findings here.

   _____

   _____

   _____

2. Squeeze each hose firmly. Do you notice any cracks or signs of splits when hoses are squeezed? Do any hoses feel mushy or extremely brittle?

   _____

   _____

   _____

3. Carefully examine the areas around the clamps. Are there any rust stains? What is or would be indicated by rust stains?

   _____

   _____

   _____

4. Carefully check the lower radiator hose. This hose contains a coiled wire lining to keep it from collapsing during operation. If the wire loses tension, the hose can partially collapse at high speed and restrict coolant flow. Describe this hose's condition.

   _____

   _____

5. Identify all hoses that need to be replaced.

   _____

   _____

   _____

   _____

6. Gather the new hoses, hose clamps, and coolant. ☐ Task completed

7. Release the pressure in the system by slowly and carefully opening the radiator cap. ☐ Task completed

8. Once the pressure is relieved, drain the coolant system below the level that is being worked on. Make sure the drained coolant is collected and recycled or disposed of according to local regulations. What did you do with the drained coolant?

   _____

   _____

   _____

   _____

9. Use a knife to cut off the old hose and loosen or cut the old clamp. ☐ Task completed

10. Slide the old hose off the fitting. If the hose is stuck, do not pry it off; instead, cut it off. ☐ Task completed

11. Clean off any remaining hose particles with a wire brush or emery cloth. What can happen if dirt or metal burrs are on the fitting when a new hose is attached?

    _____

    _____

12. Coat the surface with a sealing compound. What compound did you use?

_____

13. Place the new clamps on each end of the hose before positioning the hose.   ☐ Task completed

14. Slide the clamps to about 1/4 inch from the end of the hose after it is properly positioned on the fitting.   ☐ Task completed

15. Tighten the clamp securely. Do not overtighten.   ☐ Task completed

16. Refill the cooling system with coolant.   ☐ Task completed

17. Run the engine until it is warm.   ☐ Task completed

18. Bleed the system according to the manufacturer's recommendations. What   ☐ Task completed
    did you do to bleed the system?

_____

_____

_____

_____

19. Turn off the engine and recheck the coolant level.   ☐ Task completed

20. Retighten the heater and cooling system hose clamps.   ☐ Task completed

Instructor's Comments

_____

_____

_____

HEATING AND AIR CONDITIONING JOB SHEET 13

Servicing a Thermostat

Name _____ Station _____ Date _____

NATEF Correlation

This Job Sheet addresses the following NATEF task:

C.4. Inspect, test, and replace thermostat and housing.

Objective

Upon completion of this job sheet, you will be able to inspect, test, and replace a thermostat and housing.

Tools and Materials

Thermometer

Container for water

Heat source for heating the water

Basic hand tools

Protective Clothing

Goggles or safety glasses with side shields

Describe the vehicle being worked on:

Year _____ Make _____ Model _____

VIN _____ Engine type and size _____

PROCEDURE

1. Thoroughly inspect the area around the thermostat and its housing. Describe the result of that inspection.

   _____

   _____

   _____

2. There are several ways to test the opening temperature of a thermostat. The first method does not require that the thermostat be removed from the engine. Begin by removing the radiator pressure cap from a cool radiator and inserting a thermometer into the coolant.   ☐ Task completed

3. Start the engine and let it warm up. Watch the thermometer and the surface of the coolant. When the coolant begins to flow or move in the radiator, what is indicated?

   _____

   _____

4. When the fluid begins to flow, record the reading on the thermometer. If the engine is cold and coolant circulates, this means that the thermostat is stuck open and must be replaced. The measured opening temperature of the thermostat was _____.

5. Compare the measured opening temperature with the specifications. The specifications call for the thermostat to open at what temperature? _____ What do you recommend based on the results of this test?

   _____

   _____

6. The other way to test a thermostat is to remove it. If the cap is still on, begin removal by opening the radiator pressure cap to relieve pressure.   □ Task completed

7. Drain coolant from the radiator until the level of coolant is below the thermostat housing. Recycle the coolant according to local regulations. What did you do with the drained coolant?

   _____

   _____

   _____

8. Disconnect the upper radiator hose from the thermostat housing.   □ Task completed

9. Unbolt and remove the housing from the engine. The thermostat may come off with the housing.   □ Task completed

10. Thoroughly clean the gasket surfaces for the housing. Make sure the surfaces are not damaged while doing so and that gasket pieces do not fall into the engine at the thermostat bore.   □ Task completed

11. Carefully inspect the thermostat housing. Describe your findings.

    _____

    _____

12. Suspend the thermostat while it is completely submerged in a small container of water. Make sure it does not touch the bottom of the container. What did you use to suspend it?

    _____

    _____

    _____

13. Place a thermometer in the water so that it does not touch the container and only measures water temperature.   □ Task completed

14. Heat the water. When the thermostat valve barely begins to open, read the thermometer. What was the measured opening temperature of the thermostat? Compare that with the specifications.

    _____

    _____

15. Remove the thermostat from the water and observe the valve. Record what happened, as well as your conclusions about the thermostat.

_____

_____

_____

16. Carefully look over the new thermostat (or the old one if it is still usable). Identify which end of the thermostat should face the radiator. How did you determine the proper direction for installation?

_____

_____

17. Fit the thermostat in the recessed area in the engine or thermostat housing. Where was the recess?

_____

18. Refer to the installation instructions on the gasket's container. Should an adhesive or sealant, or both, be used with the gasket? If so, which should be used?

_____

_____

19. Install the gasket according to the instructions.   ☐ Task completed

20. Install the thermostat housing. Before tightening this into place, make sure   ☐ Task completed
    it is fully seated and flush onto the engine. Failure to do this will result in
    a broken housing.

21. Tighten the bolts evenly and carefully and to the correct specifications. What are the specifications?

_____

22. Connect the upper radiator hose to the housing with a new clamp. Why should you use a new clamp?

_____

_____

_____

23. Pressurize the system and check for leaks. Why should you do this now?

_____

_____

24. Replenish the coolant and bring it to its proper level. Install the radiator   ☐ Task completed
    cap.

25. Run the engine until it is at normal operating temperature. Did the cooling system warm up properly and does it seem the thermostat is working properly?

_____

_____

26. Recheck the coolant level.   ☐ Task completed

Instructor's Comments

_____

_____

_____

HEATING AND AIR CONDITIONING JOB SHEET 14

Checking a Mechanical Cooling Fan

Name _____ Station _____ Date _____

NATEF Correlation

This Job Sheet addresses the following NATEF task:

C.7. Inspect and test cooling fan, fan clutch, fan shroud, and air dams; perform necessary action.

Objective

Upon completion of this job sheet, you will be able to inspect and test a cooling fan, fan clutch, fan shroud, and air dams.

Tools and Materials

Vehicle with a mechanical cooling fan
Thermometer
Ignition timing light
Tachometer

Protective Clothing

Goggles or safety glasses with side shields

Describe the vehicle being worked on:

Year _____ Make _____ Model _____

VIN _____ Engine type and size _____

PROCEDURE

1. Inspect the fan shroud and air dams. Make sure they are firmly mounted to the radiator and/or frame of the vehicle. Summarize your findings.

   _____

   _____

2. Check the straightness of the fan blades. Summarize your findings.

   _____

   _____

3. Check for missing or damaged attaching rivets. Check the attaching point for each fan blade. Summarize your findings.

   _____

   _____

4. Check the condition of the drive pulley for the fan. Summarize your findings.

_____

_____

5. If the fan is equipped with a clutch, continue this job sheet.   ☐ **Task completed**

6. Use your hands to spin the fan. If the fan rotates more than twice, the clutch is bad. Summarize your findings.

_____

_____

7. Check for oil around the clutch. Oil indicates the viscous clutch is leaking and should be replaced. Summarize your findings.

_____

_____

8. Attach a thermometer to the engine side of the radiator.   ☐ **Task completed**

9. Connect a timing light and tachometer to the engine.   ☐ **Task completed**

10. Start the engine, then observe and record the reading on the thermometer   ☐ **Task completed**
when the engine is cold.

11. Aim the timing light at the fan blades; they should appear to being moving very slowly. Summarize your findings.

_____

_____

12. Put a large piece of cardboard across the radiator to block air flow. Watch the thermometer and the timing light flashes as the engine warms up. Summarize your findings.

_____

_____

13. Refer to the service manual to find the specified fan engagement temperature. What is it?

_____

_____

14. Once the engine reaches that temperature, remove the cardboard. The fan clutch should now cause the fan speed to increase. Summarize your findings.

_____

_____

Instructor's Comments

HEATING AND AIR CONDITIONING JOB SHEET 15

Clean, Inspect, Test, and Replace Electric Cooling Fans and Cooling System-Related Temperature Sensors

Name _____ Station _____ Date _____

NATEF Correlation

This Job Sheet addresses the following NATEF task:

C.8. Inspect and test electrical cooling fan, fan control system, and circuits; determine necessary action.

Objective

Upon completion of this job sheet, you will be able to demonstrate the ability to inspect, test, clean, and replace electric cooling fans and cooling system-related temperature sensors.

Tools and Materials

Hand tools

Test light (circuit tester)

Digital multimeter (DMM)

Jumper wire

Service manual

Protective Clothing

Goggles or safety glasses with side shields

Describe the vehicle being worked on:

Year _____ Make _____ Model _____

VIN _____ Engine type and size _____

Describe general condition:

_____

PROCEDURE

1. Visually inspect the electric cooling fan(s) and related wiring to determine if they are clean and properly connected.   ☐ Task completed

2. Run the engine until it reaches normal operating temperature to determine if the electric cooling fan is working.   ☐ Task completed

3. If the fan is operating properly, blow the fan and surrounding area clean with air.   ☐ Task completed

4. If the fan is not working, check the power wire to the fan to determine if there is power.   ☐ Task completed

5. Check the ground circuit to determine if the system has a good ground. ☐ Task completed

6. If power and a good ground are present, remove the electric fan assembly and replace it. ☐ Task completed

7. If there is no power, check the fuse and fan relay. The location of the fuse and relay can be found in the service manual. ☐ Task completed

8. Replace bad components, then retest fan operation. ☐ Task completed

9. If there is not a good ground, check the temperature sensor that controls the electric fan. The location of the sensor and the proper test procedure can be found in the service manual. ☐ Task completed

10. If the sensor fails the test, replace it. Check the operation of the electric fan. ☐ Task completed

   WARNING: *The electric cooling fans can operate with the engine and the key both turned off. The fan can come on at any time if the engine is hot or there is a problem with a sensor. Keep your hands away from the fan at all times.*

Problems Encountered

_____

_____

_____

Instructor's Comments

_____

_____

_____

HEATING AND AIR CONDITIONING JOB SHEET 16

Testing Heater Control Valves

Name _____ Station _____ Date _____

NATEF Correlation

This Job Sheet addresses the following NATEF task:

C.9. Inspect and test heater control valve(s); perform necessary action.

Objective

Upon completion of this job sheet, you will be able to inspect and test heater control valves properly.

Tools and Materials

Hand-operated vacuum pump

Protective Clothing

Goggles or safety glasses with side shields

Describe the vehicle being worked on:

Year _____ Make _____ Model _____

VIN _____ Engine type and size _____

PROCEDURE

1. Locate the heater control valve and inspect the hoses' connections and their clamps. Summarize your findings.

   _____

   _____

2. If the valve is mechanically controlled, check the condition of the linkage and the cable. Summarize your findings.

   _____

   _____

3. If the valve is vacuum controlled, check the condition of the vacuum hose and connection at the valve. Summarize your findings.

   _____

   _____

4. Disconnect the vacuum hose and connect a hand-operated vacuum pump to the valve.

5. Pump a vacuum at the valve and watch its operation; it should open   ☐ Task completed
   smoothly. Summarize your findings.

   _____

   _____

6. Refer to the service manual to determine if the valve is normally open or
   closed. Turn the heater control on the instrument panel off. Check the
   valve to see if it is where it should be. Summarize your findings.

   _____

   _____

7. Move the heater control to full heat and check the position of the valve.
   Summarize your findings.

   _____

   _____

Instructor's Comments

_____

_____

_____

HEATING AND AIR CONDITIONING JOB SHEET 17

Testing HVAC Controls

Name _____ Station _____ Date _____

NATEF Correlation

This Job Sheet addresses the following NATEF tasks:

D.1. Diagnose malfunctions in the electrical controls of heating, ventilation, and A/C (HVAC) systems; determine necessary action.

D.4. Diagnose malfunctions in the vacuum and mechanical components and controls of the heating, ventilation, and A/C (HVAC) system; determine necessary action.

D.6. Inspect and test A/C-heater control cables and linkages; perform necessary action.

Objective

Upon completion of this job sheet, you will be able to diagnose failures in the electrical, vacuum, and mechanical controls of heating, ventilation, and air conditioning systems.

Tools and Materials

Service manual

Vacuum gauge

Hand-operated vacuum pump

Protective Clothing

Goggles or safety glasses with side shields

Describe the vehicle being worked on:

Year _____ Make _____ Model _____

VIN _____ Engine type and size _____

PROCEDURE

1. Check the fuses for the compressor clutch and the blower motor. Summarize your findings.

_____

_____

2. Check the operation of the blower switch in all switch positions. Summarize your findings.

_____

_____

3. If there is a problem with the blower motor, check the circuit to determine the exact fault. Then, summarize your findings.

_____

_____

4. With the engine running, turn the A/C on and off and listen for the engagement of the compressor clutch. If the clutch makes a chattering noise, it may need adjustment or replacement. Summarize your findings.

_____

_____

5. If the clutch did not cycle with the switch, check to see if the clutch is constantly engaged or disengaged. If always engaged, the clutch is bad or there is a wire-to-wire short providing voltage to the clutch at all times. If the clutch is always disengaged, there is probably a problem with the clutch control circuit. Summarize your findings.

_____

_____

6. Check the electrical connections at the clutch. Summarize your findings.

_____

_____

7. Connect a jumper wire from the battery to the clutch. Check its operation now. If it operates quietly and smoothly, the problem is in the clutch control circuit. If it didn't operate normally when jumped, the clutch is bad. Summarize your findings.

_____

_____

8. Check the resistance of the clutch coil with an ohmmeter. Compare your measurement to specifications. Summarize your findings.

_____

_____

9. Using the service manual, identify any sensors or switches that control or limit the operation of the clutch. List all of them and summarize their condition.

_____

_____

_____

10. Identify and inspect the hoses and components for vacuum motors and doors. Look for disconnected or broken hoses, broken connectors, misrouted vacuum lines, and loose or disconnected electrical connectors at vacuum switches and controls. Summarize your findings.

_____

_____

11. Check the vacuum at the engine's intake manifold. Do this while the engine is idling. Summarize your findings.

_____

_____

12. Check the individual vacuum components. Use a hand-operated vacuum pump to check the component's activity and its ability to hold a vacuum. Summarize your findings.

_____

_____

13. Inspect all mechanical or cable linkages and controls. Make sure the cables are properly attached to the levers of the control switch. Summarize your findings.

_____

_____

14. If the cable(s) is equipped with an automatic adjuster, make sure the adjusting mechanism operates freely and is not damaged. Summarize your findings.

_____

_____

Instructor's Comments

_____

_____

_____

HEATING AND AIR CONDITIONING JOB SHEET 18

Diagnosing Motor-Driven Accessories

Name _____ Station _____ Date _____

NATEF Correlation

This Job Sheet addresses the following NATEF task:

D.2. Inspect and test A/C-heater blower, motors, resistors, switches, relays, wiring, and protection devices; perform necessary action.

Objective

Upon completion of this job sheet, you will be able to inspect and test A/C-heater blower motor circuits and components.

Tools and Materials

Wiring diagram for the vehicle

DMM

Protective Clothing

Goggles or safety glasses with side shields

Describe the vehicle being worked on:

Year _____ Make _____ Model _____

VIN _____ Engine type and size _____

PROCEDURE

NOTE: *Although this job sheet is focused on the blower motor, the procedures can easily be applied to all motor-driven accessories. Refer to the appropriate service manual and wiring diagram to identify the components of the circuit you wish to test. Then apply the same sequence and logic given in this job sheet.*

1. Refer to the service or owner's manual to identify the number of speeds at which the blower motor should operate. Also, identify whether the motor should run when the switch is in the off position or its lowest position. Describe how the blower motor should operate.

_____

_____

2. Check the wiring diagram and identify whether the blower circuit is controlled by a ground side switch or an insulated (power side) switch. State which one.

_____

_____

3. Start the engine and turn on the blower. Move the blower control to all available positions and summarize what happened in each.

_____

_____

4. Match the type of circuit control and the problem with one of the following.

The blower works at some speeds but not all.

a. Check the voltage to the blower motor at the various switch positions.
b. If the voltage doesn't change when a new position is selected, check the circuit from the switch to the resistor block.
c. If there was zero voltage in a switch position, check for an open in the resistor block.
d. Give a summary of the test results.

_____

_____

The blower doesn't work and is controlled by an insulated switch.

a. Check the fuse or circuit breaker. If it is open, check the circuit for a short.
b. Connect a jumper wire from a power source to the motor. If the motor operates, there is an open in the circuit between the fuse and the motor.
c. To identify the location of the open, connect a jumper wire across the switch. If the motor now operates, the switch is bad.
d. Connect a jumper wire across the resistor block. If the motor now operates, the resistor block is bad.
e. If the motor did not operate when the jumper wire connected it to a power source, connect the jumper wire from the motor to a known good ground. If the motor operates, the fault is in the ground circuit. If the motor doesn't operate, the motor is bad.
f. Give a summary of the test results.

_____

_____

The blower doesn't work and is controlled by a ground side switch.

a. Check the fuse or circuit breaker. If it is open, check the circuit for a short.
b. Connect a jumper wire from the motor's ground terminal to a known good ground. If the motor runs, the problem is in that circuit.
c. If the motor did not operate, check for voltage at the battery terminal of the motor. If no voltage is found, there is an open in the power feed circuit for the motor. If voltage was present, the motor is bad.
d. If the motor operated when the jumper wire was connected in step b, use an ohmmeter to check the ground connection of the switch.
e. If the ground is good, use a voltmeter to probe for voltage at any of the circuits from the resistor block to the switch. Replace the switch if there is power at that point.
f. If no voltage was available at the switch, check the circuit for an open. If there is not an open, suspect a faulty switch.
g. Give a summary of the test results.

_____

_____

The blower fan runs constantly.

a. If the motor runs when it should not, there is probably a short in the circuit. If a ground side switch controls the circuit, check for a short to ground in the control circuit. The exact problem can be isolated by disconnecting portions of the circuit until the motor stops. The short is in the part of the circuit that disconnected last.

b. If the circuit is controlled by an insulated switch, check for a wire-to-wire short. Check other circuits of the vehicle to identify what circuit is involved in this problem. That circuit will also experience a lack of control, or when that circuit is turned off, the blower motor will turn off. The exact problem can be isolated by disconnecting portions of the circuit until the motor stops. The short is in the part of the circuit that disconnected last.

c. Give a summary of the test results.

_____

_____

Instructor's Comments

_____

_____

_____

HEATING AND AIR CONDITIONING JOB SHEET 20

Checking an A/C Master Control Unit

Name _____ Station _____ Date _____

NATEF Correlation

This Job Sheet addresses the following NATEF task:

D.5. Inspect and test A/C-heater control panel assembly; determine necessary action.

Objective

Upon completion of this job sheet, you will be able to inspect and test an A/C-heater control panel assembly.

Tools and Materials

DMM

Service manual

Hand-operated vacuum pump

Protective Clothing

Goggles or safety glasses with side shields

Describe the vehicle being worked on:

Year _____ Make _____ Model _____

VIN _____ Engine type and size _____

PROCEDURE

1. Remove the A/C-heater master control from the instrument panel. ☐ Task completed

2. Inspect the connectors that go to the control unit and record your findings. Look for damage and corrosion.

   _____

   _____

3. Thoroughly inspect the connectors at the control unit. Look for terminal distortion (bends), corrosion, and other damage. Record your findings.

   _____

   _____

4. With an ohmmeter, check the continuity through the blower motor switch in all its positions. Summarize your findings.

   _____

   _____

5. Compare your results with the wiring diagram for the switch. Do you have good continuity at the correct terminals when the switch is in its various positions? _____

6. With the ohmmeter, check the resistance across the temperature control switch. Watch the meter as you sweep through the full range of the switch, and record your findings.

_____

_____

_____

7. If the mode selector switch is electrical, check for continuity across the terminals as you move the switch through the various mode selections. Record your findings.

_____

_____

_____

8. If the mode selector switch is a vacuum switch, apply a vacuum to the inlet of the switch and feel for a vacuum at the various tube connectors on the switch while you move the selector through its various positions. Record your findings.

_____

_____

_____

9. Compare your results to the information in the service manual, and summarize your recommendations for service to the master control unit.

_____

_____

_____

_____

10. Reinstall the old switch or install a new one.   ☐ Task completed

Instructor's Comments

_____

_____

_____

HEATING AND AIR CONDITIONING JOB SHEET 21

Inspecting the Air Duct System

Name _____ Station _____ Date _____

NATEF Correlation

This Job Sheet addresses the following NATEF task:

D.7. Inspect and test A/C-heater ducts, doors, hoses, cabin filters, and outlets; perform necessary action.

Objective

Upon completion of this job sheet, you will be able to inspect and test the A/C-heater ducts, doors, hoses, and outlets.

Tools and Materials
Service manual

Vacuum pump

Vacuum gauge

Hand tools

Protective Clothing
Goggles or safety glasses with side shields

Describe the vehicle being worked on:
Year _____ Make _____ Model _____

VIN _____ Engine type and size _____

PROCEDURE

1. Identify the vacuum source for the various doors and switches of the duct system and describe its location.

   _____

   _____

2. Disconnect that vacuum source.   ☐ Task completed

3. Connect the vacuum pump to the vacuum reserve tank.

4. Check for available vacuum at each of the vacuum-operated doors, motors, and switches. Summarize your findings.

   _____

   _____

5. Connect the vacuum gauge to the inlet for the defroster door motor(s) and record the vacuum available there when the master control switch is in the following positions (there are many variations in the names used for these operating modes; use the one that best describes the position you are testing).

   a. MAX _____ in. Hg
   b. NORM _____ in. Hg
   c. Bilevel _____ in. Hg
   d. VENT _____ in. Hg
   e. HEAT _____ in. Hg
   f. BLEND _____ in. Hg
   g. DEFROST _____ in. Hg
   h. OFF _____ in. Hg

6. Connect the vacuum gauge to the inlet for the A/C door motor(s) and record the vacuum available there when the master control switch is in the following positions.

   a. MAX _____ in. Hg
   b. NORM _____ in. Hg
   c. Bilevel _____ in. Hg
   d. VENT _____ in. Hg
   e. HEAT _____ in. Hg
   f. BLEND _____ in. Hg
   g. DEFROST _____ in. Hg
   h. OFF _____ in. Hg

7. Connect the vacuum gauge to the inlet for the outside (vent) door motor(s), and record the vacuum available there when the master control switch is in the following positions.

   a. MAX _____ in. Hg
   b. NORM _____ in. Hg
   c. Bilevel _____ in. Hg
   d. VENT _____ in. Hg
   e. HEAT _____ in. Hg
   f. BLEND _____ in. Hg
   g. DEFROST _____ in. Hg
   h. OFF _____ in. Hg

8. Connect the vacuum gauge to the inlet for the bilevel door motor(s) and record the vacuum available there when the master control switch is in the following positions.

   a. MAX _____ in. Hg
   b. NORM _____ in. Hg
   c. Bilevel _____ in. Hg
   d. VENT _____ in. Hg
   e. HEAT _____ in. Hg
   f. BLEND _____ in. Hg
   g. DEFROST _____ in. Hg
   h. OFF _____ in. Hg

9. Compare your findings to the information given in the service manual, and state your conclusions and recommendations for system service.

   _____

   _____

   _____

   _____

Instructor's Comments

HEATING AND AIR CONDITIONING JOB SHEET 22

Checking Automatic Climate Control Systems

Name _____ Station _____ Date _____

NATEF Correlation

This Job Sheet addresses the following NATEF task:

D.8. Check operation of automatic and semiautomatic HVAC control systems; determine necessary action.

Objective

Upon completion of this job sheet, you will be able to check the operation of automatic and semiautomatic HVAC control systems.

Tools and Materials

Scan tool

DMM

Service manual

Protective Clothing

Goggles or safety glasses with side shields

Describe the vehicle being worked on:

Year _____ Make _____ Model _____

VIN _____ Engine type and size _____

PROCEDURE

1. Refer to the service manual and determine if the automatic climate control system relies on the PCM, body control module (BCM), or a separate computer for control. What did you find?

   _____

   _____

2. If the unit is part of the PCM or BCM system, connect the scan tool and retrieve any codes that may be present. What did you find?

   _____

   _____

3. If the unit is controlled by its own computer system, use your service manual and record the procedure for retrieving trouble codes.

   _____

   _____

   _____

4. If there were trouble codes, what problem areas were identified?

_____

_____

5. Test the components or subsystems identified by the trouble codes, and summarize the results of those tests.

_____

_____

6. If the system is a semiautomatic system, check the service manual for the proper diagnostic procedures for checking evaporator and heating controls. Record these procedures here.

_____

_____

_____

7. Follow the appropriate test procedures and summarize your findings.

_____

_____

_____

_____

_____

Instructor's Comments

_____

_____

_____

HEATING AND AIR CONDITIONING JOB SHEET 23

Checking the Operation of Refrigerant Handling Equipment

Name _____ Station _____ Date _____

NATEF Correlation

This Job Sheet addresses the following NATEF task:

E.1. Verify correct use and maintenance of refrigerant handling equipment.

Objective

Upon completion of this job sheet, you will be able to verify correct operation and maintenance of refrigerant handling equipment.

Tools and Materials

Manifold gauge sets

Refrigerant storage containers

Refrigerant recovery machines

Protective Clothing

Goggles or safety glasses with side shields

PROCEDURE

1. Check the manifold gauge set to make sure it has a cut off valve at the end of each hose so that the fitting not in use is automatically shut. Summarize your findings.

_____

2. Check to make sure the manifold gauge set is clearly labeled as to what refrigerant it should be used for. Manifold gauge sets for R-134a can be identified by one or all of the following: Labeled FOR USE WITH R-134a, Labeled HFC-134 or R-134a, and/or have a light blue color on the face of the gauges. Summarize your findings.

_____

3. Verify that the R-134a service hoses have a black stripe along their length and are clearly labeled SAE J2196/R-134a. The low pressure hose is blue with a black stripe, the high pressure hose is red with a black stripe, and the center service hose is yellow with a black stripe. Summarize your findings.

_____

_____

4. Check the fittings at the end of the service hoses. By law, the service hoses for one type of refriger- ant should not easily connect into the wrong system, as the fittings for an R-134a system are differ- ent from those used in an R-12 system. Summarize your findings.

_____

_____

5. Make sure all A/C equipment is designated as to what refrigerant it is designed for. Because R-134a is not interchangeable with R-12, separate sets of hoses, gauges, and other equipment are required to service vehicles. All equipment used to service R-134a and R-12 systems must meet SAE standard J1991. Summarize your findings.

_____

_____

6. Make sure the containers of refrigerant are clearly labeled. When you are not sure of the refrigerant stored in a container or if you suspect a mixing of refrigerants has occurred, you should run a puri- ty test and/or use a refrigerant identifier. Summarize your findings.

_____

_____

7. Check the refrigerant storage containers. All recycled refrigerant must be safely stored in DOT CFR Title 49 or UL-approved containers. Containers specifically made for R-134a should be so marked. Before any container of recycled refrigerant can be used, it must be checked for noncondensable gases. Summarize your findings.

_____

_____

8. Locate the UL approved label on the refrigerant recovery machines. Summarize your findings.

_____

_____

Instructor's Comments

_____

_____

_____

HEATING AND AIR CONDITIONING JOB SHEET 25

Recycle, Label, and Store Refrigerant

Name _____ Station _____ Date _____

NATEF Correlation

This Job Sheet addresses the following NATEF tasks:

E.3. Recycle refrigerant.

E.4. Label and store refrigerant.

Objective

Upon completion of this job sheet, you will be able to recycle, label, and store refrigerant safely and properly.

Tools and Materials: Recycling machine, Service manual

Protective Clothing: Goggles or safety glasses with side shields

Describe the vehicle being worked on:

Year _____ Make _____ Model _____

VIN _____ Engine type and size _____

Describe the type of air conditioning found on the vehicle:

_____

PROCEDURE

1. Refer to the manual on your specific machine to determine the procedure   ☐ Task completed
   to recycle refrigerant.

2. Isolate recycling machine from the vehicle system.   ☐ Task completed

3. Recycle the refrigerant.   ☐ Task completed

4. Label the container as "empty" or "empty, evacuated and ready for dis-   ☐ Task completed
   posal" or "freon" or "contains recycled freon."

5. After the freon container has been identified, it can be stored or disposed   ☐ Task completed
   of accordingly.

Problems Encountered

_____

_____

_____

Instructor's Comments

_____

_____

_____

HEATING AND AIR CONDITIONING JOB SHEET 26

Testing for Noncondensable Gases

Name _____ Station _____ Date _____

NATEF Correlation

This Job Sheet addresses the following NATEF task:

E.5. Test recycled refrigerant for noncondensable gases.

Objective

Upon completion of this job sheet, you will be able to test recycled refrigerant for noncondensable gases according to the guidelines set by the SAE.

Tools and Materials

Pressure gauge set

Thermometer

SAE pressure/temperature chart

Protective Clothing

Goggles or safety glasses with side shields

Describe the vehicle being worked on:

Year _____ Make _____ Model _____

VIN _____ Engine type and size _____

PROCEDURE

1. To determine if the recycled refrigerant container has excess non-condensable gases, the container must be stored at a temperature of 65°F (18.3°C) or above for 12 hours and protected from the sun. ☐ Task completed

2. Install a calibrated pressure gauge onto the container and measure the pressure in the container. What pressure was read?

   _____

3. Measure the temperature of the air 4 inches away from the container's surface. What was your measurement?

   _____

4. Compare the measured pressure and temperature to the pressure/ temperature chart. Determine if the recycled refrigerant has excessive noncondensable gases. Explain your findings.

_____

_____

_____

5. If the refrigerant has excessive noncondensable gases, it should be recycled again and then retested before it is used.   ☐ Task completed

6. Summarize the results of this test.

_____

_____

_____

Instructor's Comments

_____

_____

_____

NOTICE: SOME PARTS OF THIS COPY ARE DIFFERENT THAN THE PREVIOUS PRINTING OF THIS BOOK.

The contents of this book have been updated in response to the recent changes made by NATEF. Most of these changes were minor and involved the rewording of task statements. There were, however, some new tasks added to their list. The additions resulted in the renumbering of the tasks, those changes have also been made to this workbook.

The NATEF task list included here shows where the additions, deletions, and changes were made. There is also a table that shows which Job Sheet correlates to the new NATEF task numbers, as well as the old.

Job Sheets that relate to the new tasks follow the job sheet correlation chart.

NATEF TASK LIST FOR HEATING AND AIR CONDITIONING

Legend: everything that is **new** is <u>underlined</u>
everything that has been **deleted** is ~~struck through~~

A. A/C System Diagnosis and Repair

<u>A.1.</u>	<u>Identify and interpret heating and air conditioning concern; determine necessary action.</u>	<u>Priority Rating 1</u>
<u>A.2.</u>	<u>Research applicable vehicle and service information, such as heating and air conditioning operation, vehicle service history, service precautions, and technical service bulletins.</u>	<u>Priority Rating 1</u>
<u>A.3.</u>	<u>Locate and interpret vehicle and major component identification numbers (VIN, vehicle certification labels, calibration labels).</u>	<u>Priority Rating 1</u>
<u>A.4.</u>	<u>Performance test A/C system; diagnose A/C system malfunctions using principles of refrigeration.</u>	<u>Priority Rating 1</u>
A.~~1~~.<u>5.</u>	Diagnose ~~unusual~~ <u>abnormal</u> operating noises in the A/C system; determine necessary action.	Priority Rating 2
A.~~2~~.<u>6.</u>	Identify refrigerant type; conduct a performance test of the A/C system; determine necessary action.	Priority Rating 1
A.~~3~~.<u>7.</u>	Leak test A/C system; determine necessary action.	Priority Rating 1
A.~~4~~.<u>8.</u>	Inspect the condition of discharged oil; determine necessary action.	Priority Rating 2
A.~~5~~.<u>9.</u>	~~Select oil type, measure, and add oil to the A/C system as needed.~~ <u>Determine recommended oil for system application.</u>	Priority Rating 2

B. Refrigeration System Component Diagnosis and Repair

1. Compressor and Clutch

B.1.1.	Diagnose A/C system conditions that cause the protection devices (pressure, thermal, and PCM) to interrupt system operation; determine necessary action.	Priority Rating 2
B.1.2.	Inspect A/C compressor drive belts; ~~replace and adjust as needed~~ <u>determine necessary action.</u>	Priority Rating 2
B.1.3.	Inspect, test, and<u>/or</u> replace A/C compressor clutch components and/or assembly.	Priority Rating 2
B.1.4.	Remove and ~~replace~~ <u>reinstall</u> A/C compressor and mountings; <u>measure oil quantity;</u> determine necessary action.	Priority Rating 2

2. *Evaporator, Condenser, and Related Components*

B.2.1.	Determine need for <u>an additional</u> A/C system filter; perform necessary action.	Priority Rating <u>1</u>
B.2.2.	Remove and inspect AC system mufflers, hoses, lines, fittings, o-rings, seals, and service valves; perform necessary action.	Priority Rating 2
B.2.3.	Inspect A/C condenser for airflow restrictions; perform necessary action.	Priority Rating 1
B.2.4.	Remove and install receiver/drier or accumulator/drier; <u>measure oil quantity; determine necessary action</u>.	Priority Rating 2
B.2.5.	Remove and install expansion valve or orifice (expansion) tube.	Priority Rating 2
B.2.6.	Inspect evaporator housing water drain; perform necessary action.	Priority Rating 3
<u>B.2.7.</u>	<u>Remove and reinstall evaporator; measure oil quantity; determine necessary action.</u>	<u>Priority Rating 1</u>
<u>B.2.8.</u>	<u>Remove and reinstall condenser; measure oil quantity; determine necessary action.</u>	<u>Priority Rating 1</u>

C. Heating, Ventilation, and Engine Cooling Systems Diagnosis and Repair

C.1.	Diagnose temperature control problems in the heater/ventilation system; determine necessary action.	Priority Rating 2
C.2.	Perform cooling system, cap, and recovery system tests (pressure, combustion leakage, and temperature); determine necessary action.	Priority Rating 1
C.3.	Inspect engine cooling and heater system hoses and belts; perform necessary action.	Priority Rating 1
C.4.	Inspect, test, and replace thermostat and housing.	Priority Rating 1
C.5.	Determine coolant condition <u>and coolant type for vehicle application</u>; drain and recover coolant.	Priority Rating 1
C.6.	Flush system; refill system with recommended coolant; bleed system.	Priority Rating 1
C.7.	Inspect and test <u>cooling </u>fan, fan clutch ~~(electrical and mechanical)~~, fan shroud, and air dams; perform necessary action.	Priority Rating 1
C.8.	Inspect and test electrical <u>cooling fan,</u> fan control system, and circuits<u>; determine necessary action</u>.	Priority Rating 1
C.9.	Inspect and test heater control valve(s); perform necessary action.	Priority Rating 2
<u>C.10.</u>	<u>Remove and reinstall heater core.</u>	<u>Priority Rating 1</u>

D. Operating Systems and Related Controls Diagnosis and Repair

D.1.	Diagnose ~~failures~~ <u>malfunctions </u>in the electrical controls of heating, ventilation, and A/C (HVAC) systems; determine necessary action.	Priority Rating 2
D.2.	Inspect and test A/C-heater blower, motors, resistors, switches, relays, wiring, and protection devices; perform necessary action.	Priority Rating 2
D.3.	Test <u>and diagnose</u> A/C compressor ~~load cut-off~~ <u>clutch control</u> systems; determine necessary action.	Priority Rating 3
D.4.	Diagnose ~~failures~~ <u>malfunctions </u>in the vacuum and mechanical components and controls of the <u>heating, ventilation, and A/C (HVAC)</u> system; determine necessary action.	Priority Rating 2
D.5.	Inspect and test A/C-heater control panel assembly; determine necessary action.	Priority Rating 3

D.6.   Inspect and test A/C-heater control cables and linkages; perform necessary action.   Priority Rating 3

D.7.   Inspect and test A/C-heater ducts, doors, hoses, <u>cabin filters,</u> and outlets; perform necessary action.   Priority Rating 3

D.8.   Check operation of automatic and semiautomatic <u>heating, ventilation, and air-conditioning (HVAC)</u> control systems; determine necessary action.   Priority Rating 3

E.   Refrigerant Recovery, Recycling, and Handling

E.1.   Verify correct ~~operation~~ <u>use</u> and maintenance of refrigerant handling equipment.   Priority Rating 1

E.2.   Identify (by label application or use of a refrigerant identifier) and recover A/C system refrigerant.   Priority Rating 1

E.3.   Recycle refrigerant.   Priority Rating 1

E.4.   Label and store refrigerant.   Priority Rating 1

E.5.   Test recycled refrigerant for noncondensable gases.   Priority Rating 1

E.6.   Evacuate and charge A/C system.   Priority Rating 1

New Task #	Old Task #	Job Sheet #
A.1	NEW	28
A.2	NEW	29
A.3	NEW	29
A.4	NEW	30
A.5	A.1	1
A.6	A.2	2
A.7	A.3	3
A.8	A.4	4
A.9	A.5	4
B.1.1	B.1.1	5
B.1.2	B.1.2	1
B.1.3	B.1.3	5
B.1.4	B.1.4	6
B.2.1	B.2.1	7
B.2.2	B.2.2	8
B.2.3	B.2.3	1
B.2.4	B.2.4	8
B.2.5	B.2.5	8
B.2.6	B.2.6	9
B.2.7	NEW	31
B.2.8	NEW	32
C.1	C.1	10
C.2	C.2	11
C.3	C.3	12
C.4	C.4	13
C.5	C.5	11
C.6	C.6	11
C.7	C.7	14
C.8	C.8	15
C.9	C.9	16
C.10	NEW	33
D.1	D.1	17
D.2	D.2	18
D.3	D.3	19
D.4	D.4	17
D.5	D.5	20
D.6	D.6	17
D.7	D.6	21
D.8	D.8	22
E.1	E.1	23
E.2	E.1	24
E.3	E.3	25
E.4	E.4	25
E.5	E.5	26
E.6	E.6	27

HEATING AND AIR CONDITIONING JOB SHEET 28

Identifying Problems and Concerns

Name _____ Station _____ Date _____

NATEF Correlation

This Job Sheet addresses the following NATEF task:

A.1. Identify and interpret heating and air conditioning concern; determine necessary action.

Objective

Upon completion of this job sheet, you will be able to define heating and air conditioning system problems or concerns, prior to diagnosing or testing the systems.

Protective Clothing

Goggles or safety glasses with side shields

Describe the vehicle being worked on:

Year _____ Make _____ Model _____

VIN _____ Engine type and size _____

PROCEDURE

1. Start the engine and describe how the engine seems to be running:

   _____

   _____

2. Move the temperature control on the dash to its full heat position. Describe the amount of heat coming from the heater in each of the fan positions.

   _____

   _____

3. Were there any unusual noises or smells when the heater fan was running? If so, describe them and when they occurred.

   _____

   _____

4. Turn the temperature control to the defrost mode. Describe the amount of heat coming from the defroster in each of the fan positions.

   _____

   _____

5. Were there any unusual noises or smells when the heater fan was running? If so, describe them and when they occurred.

_____

_____

6. Turn the temperature control to cool or the air conditioning position. Describe what the engine did when the A/C was turned on.

_____

_____

7. Describe how the air feels in each of the various fan positions.

_____

_____

8. Were there any unusual noises or smells when the A/C fan was running? If so, describe them and when they occurred.

_____

_____

9. Summarize your findings in detail; include both improper and proper system operation.

_____

_____

_____

_____

_____

_____

_____

10. Based on the above, what are your suspicions and conclusions?

_____

_____

_____

_____

Instructor's Comments

_____

_____

_____

HEATING AND AIR CONDITIONING JOB SHEET 29

Gathering Vehicle Information

Name _____ Station _____ Date _____

NATEF Correlation

This Job Sheet addresses the following NATEF tasks:

A.2. Research applicable vehicle and service information, such as heating and air conditioning operation, vehicle service history, service precautions, and technical service bulletins.

A.3. Locate and interpret vehicle and major component identification numbers (VIN, vehicle certification labels, calibration labels).

Objective

Upon completion of this job sheet, you will be able to gather service information about a vehicle and its heating and air conditioning system.

Tools and Materials

Appropriate service manuals
Computer

Protective Clothing

Goggles or safety glasses with side shields

Describe the vehicle being worked on:

Year _____ Make _____ Model _____

VIN _____

PROCEDURE

1. Using the service manual or other information source, describe what each letter and number in the VIN for this vehicle represents.

   _____

   _____

2. Locate the Vehicle Emissions Control Information (VECI) label and describe where you found it.

   _____

   _____

3. Summarize what information you found on the VECI label.

   _____

   _____

4. While looking in the engine compartment did you find a label regarding the vehicle's air conditioning system? Describe where you found it.

_____

_____

5. Summarize the information contained on this label.

_____

_____

6. Using a service manual or electronic database, locate the information about the vehicle's air conditioning system. List the major components of the system and describe how the system's pressure is controlled.

_____

_____

_____

_____

_____

7. Using a service manual or electronic database, locate and record all service precautions regarding the air conditioning system noted by the manufacturer.

_____

_____

_____

_____

8. Using the information that is available, locate and record the vehicle's service history.

_____

_____

_____

_____

9. Using the information sources that are available, summarize all Technical Service Bulletins for this vehicle that relate to the heating and air conditioning system.

_____

_____

_____

_____

Instructor's Comments

_____

_____

_____

HEATING AND AIR CONDITIONING JOB SHEET 30

Conducting a Performance Test on an Air Conditioning System

Name _____ Station _____ Date _____

NATEF Correlation

This Job Sheet addresses the following NATEF task:

A.4. Performance test A/C system; diagnose A/C system malfunctions using principles of refrigeration.

Objective

Upon completion of this job sheet, you will be able to conduct an A/C system performance test and use theory to help diagnose basic problems within the system.

Tools and Materials
Service manual
Thermometer

Protective Clothing
Goggles or safety glasses with side shields

Describe the vehicle being worked on:

Year _____ Make _____ Model _____

VIN _____ Engine type and size _____

PROCEDURE

1. Describe the type of air conditioning system that is on this vehicle:

   _____

   _____

   _____

   _____

2. Carefully inspect the condenser for damage and check for anything that may block air flow. Describe your findings:

   _____

   _____

3. Inspect the drive belts and hoses for the A/C system and describe your findings:

   _____

   _____

4. Check the mounting of the compressor and describe your findings.

_____

_____

5. Start the engine and operate the heating and air conditioning control in each of its positions. List any switch position that the system did not respond to correctly:

_____

_____

6. Keeping your hands away from the cooling fan, measure the temperature of the air in front of the condenser. What was your measurement?

_____

7. Place the thermometer in the center duct in the dash and set the A/C con-   ☐ Task completed
   trol to its maximum position and the fan to its highest speed.

8. Raise the speed of the engine to about 1500 rpm and note the reading on the thermometer. What was it?

_____

9. What reading would be satisfactory? Why? Explain why outside ambient temperature will affect the temperature of the air coming out of the duct.

_____

_____

_____

_____

10. Listen to the system and record any unusual sounds. Be sure to listen to the system while the compressor clutch cycles on and off.

_____

_____

11. Carefully place your hand on the condenser near the refrigerant inlet. Describe what you feel:

_____

_____

12. Carefully place your hand on the condenser near the refrigerant outlet. Describe what you feel:

_____

_____

13. Was there a change in temperature between the inlet and the outlet?

    _____

14. Based on the above inspection and performance test, what are your conclusions and recommendations?

    _____

    _____

    _____

    _____

Instructor's Comments

_____

_____

_____

HEATING AND AIR CONDITIONING JOB SHEET 31

Removing and Installing an Evaporator

Name _____ Station _____ Date _____

NATEF Correlation

This Job Sheet addresses the following NATEF task:

B.2.7. Remove and reinstall evaporator; measure oil quantity; determine necessary action.

Objective

Upon completion of this job sheet, you will be able to correctly remove and reinstall an evaporator. You will also be able to determine the amount of oil that should be added to the system after a new evaporator has been installed.

Tools and Materials

Charging station	Miscellaneous plugs and hoses
Gauge set	New refrigerant oil
Graduated container	Refrigerant recovery/recycling machine
Hand tools	Service manual

Protective Clothing

Goggles or safety glasses with side shields

Describe the vehicle being worked on:

Year _____ Make _____ Model _____

VIN _____ Engine type and size _____

PROCEDURE

1. Recover the refrigerant using an approved recovery/recycling equipment. ☐ Task completed

2. Disconnect the negative battery cable. ☐ Task completed

3. If the evaporator and heater core are a combined unit, drain the engine's coolant. ☐ Task completed

4. Disconnect and label all electrical connectors, cables, and vacuum hoses that are connected to the evaporator. What did you need to disconnect?

_____

_____

5. Disconnect the refrigerant hoses at the evaporator and plug or cap the hose ends to prevent dirt and moisture from entering the system. ☐ Task completed

6. Unbolt and remove the evaporator. What did you need to remove to gain access to the evaporator?

_____

_____

7. Drain the oil from the evaporator into a graduated container. How much oil was drained from the evaporator?

_____

8. Check the oil for dirt. If the oil is contaminated, replace the accumulator or receiver/drier. Describe the condition of the oil.

_____

_____

9. Add the above amount of new refrigerant oil or the amount specified in the service manual to the new evaporator. How much oil did you put into the evaporator?

_____

_____

10. Coat the new O-rings for the evaporator and the line fittings on the evaporator with clean refrigerant oil.   ☐ Task completed

11. Install the new evaporator and tighten the fittings.   ☐ Task completed

12. Reinstall and reconnect all parts, wires, cables, and vacuum hoses that were disconnected during removal. Add coolant if needed.   ☐ Task completed

13. Connect the negative battery cable.   ☐ Task completed

14. Evacuate and recharge the system. How much refrigerant was put in the system?

_____

15. Perform a leak test and correct any problems. Were there any problems?

_____

_____

16. Drive the vehicle and then conduct a performance test. Does the system work as it should? Explain.

_____

_____

_____

_____

Instructor's Comments

_____

_____

_____

HEATING AND AIR CONDITIONING JOB SHEET 32

Removing and Installing a Condenser

Name _____ Station _____ Date _____

NATEF Correlation

This Job Sheet addresses the following NATEF task:

B.2.8. Remove and reinstall condenser; measure oil quantity; determine necessary action.

Objective

Upon completion of this job sheet, you will be able to correctly remove and reinstall a condenser. You will also be able to determine the amount of oil that should be added to the system after a new condenser has been installed.

Tools and Materials

Charging station	Miscellaneous plugs and hoses
Gauge set	New refrigerant oil
Graduated container	Refrigerant recovery/recycling machine
Hand tools	Service manual

Protective Clothing

Goggles or safety glasses with side shields

Describe the vehicle being worked on:

Year _____ Make _____ Model _____

VIN _____ Engine type and size _____

PROCEDURE

1. Recover the refrigerant using an approved recovery/recycling equipment.   ☐ Task completed

2. Disconnect the negative battery cable.   ☐ Task completed

3. On many vehicles, the radiator, cooling fan, and/or shroud must be removed to gain access to the condenser. If the radiator must be removed, drain the coolant. What do you need to remove?

   _____

   _____

4. Disconnect and label all electrical connectors that are connected to the condenser and all those that may be in the way. What did you need to disconnect?

   _____

   _____

5. Disconnect the refrigerant hoses at the condenser and plug or cap the hose ends to prevent dirt and moisture from entering the system.   ☐ Task completed

6. Unbolt and remove the evaporator. If the receiver/drier mounts to the condenser, disconnect it and remove it with the condenser.   ☐ Task completed

7. Drain the oil from the condenser into a graduated container. How much oil was drained from the condenser?

   _____

8. Check the oil for dirt. If the oil is contaminated, replace the accumulator or receiver/drier. Describe the condition of the oil.

   _____

   _____

9. Add the above amount of new refrigerant oil or the amount specified in the service manual to the new condenser. How much oil did you put into the evaporator?

   _____

   _____

10. Install the new condenser and loosely tighten the mounting bolts.   ☐ Task completed

11. Coat the new O-rings for the condenser and the line fittings on the condenser with clean refrigerant oil.   ☐ Task completed

12. Tighten the fittings, then tighten the mounting bolts.   ☐ Task completed

13. Reinstall and reconnect all parts, wires, cables, and vacuum hoses that were disconnected or removed during removal. Add coolant if needed.   ☐ Task completed

14. Connect the negative battery cable.   ☐ Task completed

15. Evacuate and recharge the system. How much refrigerant was put in the system?

   _____

16. Perform a leak test and correct any problems. Were there any problems?

   _____

   _____

17. Drive the vehicle and then conduct a performance test. Does the system work as it should? Explain.

   _____

   _____

   _____

Instructor's Comments

_____

_____

_____

HEATING AND AIR CONDITIONING JOB SHEET 33

Removing and Installing a Heater Core

Name _____ Station _____ Date _____

NATEF Correlation

This Job Sheet addresses the following NATEF task:

C.10. Remove and reinstall heater core.

Objective

Upon completion of this job sheet, you will be able to correctly remove and reinstall a heater core.

Tools and Materials

Catch basin   Coolant recycling machine
Hand tools   Service manual

Protective Clothing

Goggles or safety glasses with side shields

Describe the vehicle being worked on:

Year _____ Make _____ Model _____

VIN _____ Engine type and size _____

PROCEDURE

1. Drain and recycle the coolant from the engine with the approved recycling equipment.   ☐ Task completed

2. Disconnect the negative battery cable.   ☐ Task completed

3. Refer to the service manual for the recommended procedure for removing the heater core. Summarize the steps here:

   _____

   _____

   _____

   _____

4. If the evaporator and heater core are a combined unit, recover and recycle the refrigerant in the A/C system.   ☐ Task completed

5. Disconnect and label all electrical connectors that are connected to the heater core and all those that may be in the way. What did you need to disconnect?

   _____

   _____

6. Loosen the hose clamps on the hoses connected to the heater core. ☐ Task completed

7. Remove the coolant hoses at the heater core and plug or cap the hose ends to prevent more lose of coolant. ☐ Task completed

8. Inspect the heater hoses and record your findings here:

   _____

   _____

9. Replace the heater hoses if necessary or desired. ☐ Task completed

10. Unbolt and remove the heater core. ☐ Task completed

11. Loosely place the new heater core into position. ☐ Task completed

12. Install new hose clamps over the hose ends. ☐ Task completed

13. Slip the hose ends into position over the inlet and outlet nipples on the heater core. ☐ Task completed

14. Tighten the hose clamps. ☐ Task completed

15. Tighten the mounting bolts for the heater core. ☐ Task completed

16. Reinstall and reconnect all parts, wires, cables, and vacuum hoses that were disconnected or removed during removal. ☐ Task completed

17. Fill the cooling system with new clean coolant and bleed the system. ☐ Task completed

18. Connect the negative battery cable. ☐ Task completed

19. Perform a leak test and correct any problems. Were there any problems?

    _____

    _____

20. Drive the vehicle and check the heating system. Does the system work as it should? Explain.

    _____

    _____

    _____

    _____

Instructor's Comments

_____

_____